American Book Company's

Passing the

Georgia

Mathematics I

End-of-Course Test

Erica Day

Colleen Pintozzi

Kali Daniel

American Book Company

P. O. Box 2638

Woodstock, Georgia 30188-1383

Toll Free 1 (888) 264-5877 Phone (770) 928-2834

Toll Free Fax 1 (866) 827-3240

WEB SITE: www.americanbookcompany.com

Acknowledgements

In preparing this book, we would like to acknowledge Mary Stoddard and Eric Field for their contributions in developing graphics for this book and Samuel Rodriguez for his contribution in editing this book. We would also like to thank our many students whose needs and questions inspired us to write this text.

Printed in the United States of America
10/09   08/09   11/08   02/08

Contents

Contents

Contents

Preface

Passing the Georgia Mathematics I End-of-Course Test will help you review and learn important concepts and skills related to algebra, geometry, data analysis, and probability. First, take the Diagnostic Test beginning on page 1 of the book. Next, complete the evaluation chart with your instructor in order to help you identify the chapters which require your careful attention. When you have finished your review of all of the material your teacher assigns, take the practice tests to evaluate your understanding of the material presented in this book. **The materials in this book are based on the Georgia Performance Standards that are published by the Georgia Department of Education. The complete list of standards is located in the Answer Key. Each question in the Diagnostic and Practice Tests is referenced to the standard, as is the beginning of each chapter.**

This book contains several sections. These sections are as follows: 1) A Diagnostic Test; 2) Chapters that teach the concepts and skills for ***Passing the Georgia Mathematics I End-of-Course Test***; and 3) Two Practice Tests. Answers to the tests and exercises are in a separate manual.

ABOUT THE AUTHORS

Erica Day has a Bachelor of Science Degree in Mathematics and is working on a Master of Science Degree in Mathematics. She graduated with high honors from Kennesaw State University in Kennesaw, Georgia. She has also tutored all levels of mathematics, ranging from high school algebra and geometry to university-level statistics, calculus, and linear algebra. She is currently writing and editing mathematics books for American Book Company, where she has coauthored numerous books, such as *Passing the Georgia Algebra I End of Course, Passing the Georgia High School Graduation Test in Mathematics, Passing the Arizona AIMS in Mathematics*, and *Passing the New Jersey HSPA in Mathematics*, to help students pass graduation and end of course exams.

Colleen Pintozzi has taught mathematics at the middle school, junior high, senior high, and adult level for 22 years. She holds a B.S. degree from Wright State University in Dayton, Ohio and has done graduate work at Wright State University, Duke University, and the University of North Carolina at Chapel Hill. She is the author of many mathematics books including such best-sellers as *Basics Made Easy: Mathematics Review, Passing the New Alabama Graduation Exam in Mathematics, Passing the Louisiana LEAP 21 GEE, Passing the Indiana ISTEP+ GQE in Mathematics, Passing the Minnesota Basic Standards Test in Mathematics*, and *Passing the Nevada High School Proficiency Exam in Mathematics*.

Kali Daniel is a student at Kennesaw State University in Kennesaw, GA en route to a Bachelor's Degree in Secondary Mathematics Education. She has also tutored students for over six years in all levels of Mathematics ranging from elementary to university-level Calculus.

Mathematics I Formula Sheet

Below are the formulas you may find useful as you work the problems. However, some of the formulas may not be used. You may refer to this page as you take the test.

Area

Rectangle and Parallelogram $A = bh$

Triangle $A = \frac{1}{2} bh$

Circle $A = \pi r^2$

Trapezoid $A = \frac{1}{2} (h)(b_1 + b_2)$

Circumference

$C = \pi d$ $\pi \approx 3.14$

Volume

Rectangular Prism/Cylinder $V = Bh$

Pyramid/Cone $V = \frac{1}{3} Bh$

Surface Area

Rectangular Prism $SA = 2lw + 2wh + 2lh$

Cylinder $SA = 2\pi r^2 + 2\pi rh$

Pythagorean Theorem

$$a^2 + b^2 = c^2$$

Mean Absolute Deviation

$$\frac{\sum_{i=1}^{N} \left| x_i - \overline{x} \right|}{N}$$

the average of the absolute deviations from the mean for a set of data

Expected Value

$$E(x) = \sum_{i=1}^{n} x_i \, p(x_i)$$

the sum of each outcome multiplied by its probability of occurrence

Permutations

$$_nP_r = \frac{n!}{(n-r)!}$$

Combinations

$$_nC_r = \frac{n!}{r!(n-r)!}$$

Interquartile Range

the difference between the first quartile and third quartile of a set of data

Diagnostic Test

Part 1

1. Which is the graph of $f(x) = x^2$?

(A)

(B)

(C)

(D)

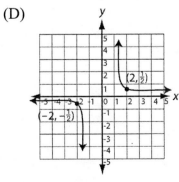

MM1A1b

2. $f(x) = 4x + 6$; What is $f(2)$?

(A) 10
(B) 12
(C) 14
(D) 16

MM1A1a

3. What is the domain of the function
$f(x) = \dfrac{1}{x}$?

(A) $(-\infty, \infty)$
(B) $(-\infty, 0) \cup (0, \infty)$
(C) $(-\infty, 1) \cup (1, \infty)$
(D) $(-\infty, -1) \cup (-1, \infty)$

MM1A1d

4. What is the pattern in the following
sequence $\{3, 4, 5, 6, 7, ...\}$?

(A) n
(B) $n + 1$
(C) $n + 2$
(D) n^2

MM1A1f

5. What is the intersection point of the graphs
of the equations $f(x) = 4x - 1$ and
$g(x) = x + 2$?

(A) $(1, 3)$
(B) $\left(\frac{1}{5}, 2\frac{1}{5}\right)$
(C) $(3, 1)$
(D) $(2, 4)$

MM1A1i

6. Subtract $y^2 + 4$ from $3y^2 + 8y$.

(A) $4y^2 + 12y$
(B) $4y^2 + 8y + 4$
(C) $2y^2 + 8y - 4$
(D) $2y^2 + 8y + 4$

MM1A2c

7. Multiply $\dfrac{x^3 - x^2}{y^2}$ and $\dfrac{y^2}{x}$.

(A) $\dfrac{x^4 - x^3}{y^2 x}$

(B) $\dfrac{x^4 y^2 - 2x^3 y^2}{y^2 x}$

(C) $\dfrac{x^3 y^2 - x^2 y^2}{y^2 x}$

(D) $x^2 - x$

MM1A2e

8. Factor the expression $x^2 + x - 30$.

(A) $(x - 6)(x + 5)$
(B) $(x - 10)(x + 3)$
(C) $(x - 2)(x + 15)$
(D) $(x + 6)(x - 5)$

MM1A2f

9. Given the following measures of the angles of a triangle, find x. $90°, 70°, x$.

(A) $10°$
(B) $20°$
(C) $30°$
(D) $40°$

MM1G3a

10. A square is _____ a rectangle.

(A) Always
(B) Sometimes
(C) Never
(D) Not enough information

MM1G3d

11. Given the following triangle, which side is the longest?

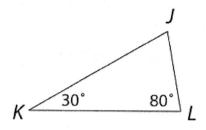

(A) $\overline{KJ}$
(B) $\overline{JL}$
(C) $\overline{KL}$
(D) Both $\overline{KJ}$ and $\overline{KL}$ are the same.

MM1G3b

12. Given the two triangles, which theorem for triangles proves they are equivalent?

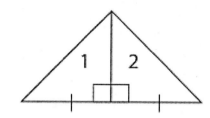

(A) SSS
(B) HL
(C) SAS
(D) ASA

MM1G3c

13. A _____ creates 6 subtriangles, which have equal areas.

(A) centroid
(B) incenter
(C) circumcenter
(D) orthocenter

MM1G3e

14. Sally is seven years old and goes to the store with her mother every Saturday morning. Every time they go, she notices that her mother gives the check out clerk money before they leave the store. Sally concluded that everyone has to give money in exchange for things they want to take out of the store. This an example of:

(A) deductive reasoning
(B) inductive reasoning
(C) analytical reasoning
(D) none of the above

MM1G2a

15. If I make all passing grades this week, then I will go to the movies on Saturday. What is the inverse of this statement?

(A) If I don't make all passing grades this week, then I won't go to the movies on Saturday.
(B) If I go to the movies on Saturday, then I made all passing grades this week.
(C) If I don't go to the movies on Saturday, then I didn't make all passing grades this week.
(D) If I don't make all passing grades this week, then I will go to the movies on Saturday.

MM1G2b

16. A = the number of math majors
B = the number present in class today
$A \cup B$ = math majors or present
$A \cap B$ = math majors and present
$n(A \cup B) = n(A) + n(B) - n(A \cap B)$
Using this information, find the probability that a student is a math major or present if $A = 36$, $B = 22$, and $A \cap B = 13$.

(A) 40
(B) 45
(C) 50
(D) 58

MM1D1a

17. How many permutations can you find using the letters in the word CAT?

(A) 3
(B) 6
(C) 9
(D) 12

MM1D1b

18. If Joe rolls 2 six-sided number cubes, the probability of him rolling a six and a double $(3, 3)$ is

(A) mutually exclusive.
(B) not mutually exclusive.
(C) both A & B.
(D) not enough information

MM1D2a

19. If Jonathan is rolling a six-sided number cube and spinning a spinner, these two events are

(A) independent.
(B) dependent.
(C) mutually exclusive.
(D) not mutually exclusive.

MM1D2b

20. Mrs. Smith was looking at her class' test scores. After two tests, 25% of her students had passed both tests. 42% passed the first test. What percent that passed the first test also passed the second?

(A) 17%
(B) 60%
(C) 67%
(D) 77%

MM1D2c

21. There are 24 candy-coated chocolate pieces in a bag. Eight have defects in the coating that can be seen only with close inspection. What is the probability of pulling out a defective piece without looking?

(A) 8

(B) $\dfrac{8}{23}$

(C) $\dfrac{1}{4}$

(D) $\dfrac{1}{3}$

MM1D2d

22. What is the mean absolute deviation of the following set of numbers $\{2, 7, 9, 1, 6, 5\}$?

(A) 1.5
(B) 2.8
(C) 2.33
(D) 3

MM1D4

23. The school cafeteria workers are conducting a survey of students' likes and dislikes for the foods they serve at lunch time. Once they tally the data and begin to order the food they need, which measure of central tendency will help most to make food purchases?

(A) Mean
(B) Mode
(C) Median
(D) Range

MM1D3a

24. $y = x^3 - 2$ is a(n) _____ function.

(A) odd
(B) even
(C) neither
(D) not enough information

MM1A1h

25. What are the zeros of $y = x^2 + 5x + 6$?

(A) $x = 3, 2$
(B) $x = -3, 2$
(C) $x = 3, -2$
(D) $x = -3, -2$

MM1A3a

26. $\sqrt{x} + 2 = 5$. What is x?

(A) 3
(B) $\sqrt{3}$
(C) 6
(D) 9

MM1A3b

27. According to the chart, what are the solution(s) (zeros) to the equation $f(x) = x^2 + 2x + 1$?

x	$f(x)$
-2	1
-1	0
0	1
1	4
2	9

(A) -1
(B) 1
(C) $-1, 1$
(D) $1, -1$

MM1A3c

28. The points $(0, 0)$, $(5, 0)$, $(0, 4)$ form what type of triangle?

(A) Isosceles
(B) Equilateral
(C) Scalene
(D) Right

MM1G1e

29. Solve for x: $3(x - 2) - 1 = 6(x + 5)$

(A) -4

(B) $-\dfrac{37}{3}$

(C) 4

(D) $\dfrac{23}{3}$

MM1A3d

30. What is the distance between $(1, 1)$ and $(6, 6)$?

(A) $\sqrt{25}$

(B) 5

(C) $\sqrt{50}$

(D) 25

MM1G1a

31. What is the shortest distance between $\overline{AC}$ and point B?

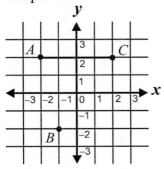

(A) 4 units
(B) 2 units
(C) 16 units
(D) 8 units

MM1G1b

32. What is the midpoint of $(6, 4)$ and $(3, -4)$?

(A) $(4.5, 0)$

(B) $(0, 4.5)$

(C) $(-4.5, 0)$

(D) $(0, -4.5)$

MM1G1c

33. Samantha is a student at Etowah High School. Look at the following histograms. The first histogram displays the number of sodas consumed each day by each student in Samantha's math class. The second histogram displays the number of sodas consumed each day by each of the students in the ninth-grade class at her school.

Find the difference in lower quartiles between the two data sets displayed in the histograms.

(A) 0

(B) 1

(C) 2

(D) 3

MM1D3b

34. Nicole is in charge of buying food for a political campaign picnic lunch in City Center Park. In order to determine which foods people would prefer, Nicole conducted a survey of food preferences in front of a small health-food store near the park. The results indicated that most people prefer fruits and vegetables. Is this conclusion valid? Use mathematics to justify your answer.

(A) The conclusion is invalid because she did not choose a random sample. In front of a health food store, there is a disproportionately high number of people who prefer health food.

(B) The conclusion is valid because she did choose a random sample.

(C) The conclusion is invalid because she did not choose a enough people for her sample.

(D) The conclusion is valid because she did survey people that are more knowledgable about food.

MM1D3c

35. Use the points $A = (0,0)$, $B = (3,2)$, $C = (0,2)$ and the Pythagorean Theorem to find the distance between A and B. What is the distance?

(A) $\sqrt{13}$
(B) 13
(C) 5
(D) $\sqrt{5}$

MM1G1d

36. Two six-sided number cubes each have faces numbered 1 through 6. The two cubes are rolled at the same time. What is the probability that the sum of the two top faces will be either a 11 or a 5?

(A) $\dfrac{1}{9}$

(B) $\dfrac{1}{18}$

(C) $\dfrac{1}{6}$

(D) $\dfrac{1}{36}$

MM1D2a

Part 2

37. Which is the graph of $f(x) = \dfrac{1}{x}$?

(A)

(B)

(C)

(D)

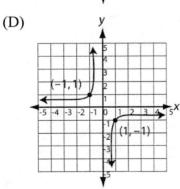

MM1A1b

38. What is the general shape of the function $y = x^2$?

(A) Straight line
(B) Parabola
(C) N shape
(D) V shape

MM1A1e

39. What is the rate of change of $f(x) = -\dfrac{3}{4}x + 2$?

(A) $-\dfrac{3}{4}$

(B) $\dfrac{3}{4}$

(C) 2

(D) -2

MM1A1g

40. What is $12\sqrt{3} - 3\sqrt{3}$?

(A) $4\sqrt{3}$
(B) $3\sqrt{3}$
(C) $9\sqrt{2}$
(D) $9\sqrt{3}$

MM1A2b

41. Multiply the binomials $(x + 2)(x + 1)$.

(A) $x^2 + 2x + 3$
(B) $x^2 + 2x + 2$
(C) $x^2 + 3x + 1$
(D) $x^2 + 3x + 2$

MM1A2d

42. What is the area of a triangle with height $h = 6 + x$ and base $b = 13 + x$?

(A) $x^2 + 19x + 78$

(B) $\dfrac{1}{2}x^2 + 19x + 39$

(C) $\dfrac{1}{2}x^2 + \dfrac{19}{2}x + 39$

(D) $\dfrac{1}{2}x^2 + \dfrac{19}{2}x + 78$

MM1A2g

43. What are the factors of $b^2 - 4b - 5$?

(A) $(b - 5)(b + 1)$
(B) $(b + 5)(b - 1)$
(C) $(b - 3)(b + 2)$
(D) $(b - 2)(b - 3)$

MM1A2f

44. Given the following triangle, find x.

(A) 30°
(B) 40°
(C) 50°
(D) 60°

MM1G3a

45. Which of the following can be the measure of the third side of a triangle if the other two sides are 7 and 13?

(A) 3
(B) 4
(C) 5
(D) 8

MM1G3b

46. Given the two triangles, which theorem for triangles proves these are equivalent?

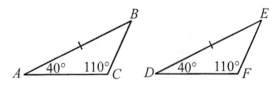

(A) SSS
(B) HL
(C) SAS
(D) AAS

MM1G3c

47. Given these two triangles, which theorem for triangles proves these are equivalent?

(A) SSS
(B) HL
(C) SAS
(D) ASA

MM1G3c

48. A rectangle is _____ a square.

(A) always
(B) sometimes
(C) never
(D) Not enough information

MM1G3d

49. The _____ is equidistant from the vertices of a triangle.

(A) centroid
(B) incenter
(C) circumcenter
(D) orthocenter

MM1G3e

50. A _____ is the center of an inscribed circle.

(A) centroid
(B) incenter
(C) circumcenter
(D) orthocenter

MM1G3e

51. All carnivores are meat eaters. Lions eat meat. Therefore, lions are carnivores. This kind of thinking is an example of _____ reasoning.

(A) applied
(B) inductive
(C) qualitative
(D) deductive

MM1G2a

52. If a figure is a triangle, then it has three sides. Which of the following statements is the contrapositive of the statement?

(A) If a figure is a triangle, then it has three sides.

(B) If a figure is not a triangle, then it does not have three sides.

(C) If a figure has three sides, then it is a triangle.

(D) If a figure does not have three sides, then it is not a triangle.

MM1G2b

53. At a toy manufacturing company in Nevada, the probability that they will make zero defective toys (out of every 100) is 0.27. The probability of making one defective toy is 0.44. What is the probability of making less than 1 defective toy?
(Hint: This is a mutually exclusive event.)

(A) 0.71

(B) 0.12

(C) 0.17

(D) 0.61

MM1D2a

54. If Jessica has a bag containing 3 blue bracelets and 7 green ones, what is the probability that she draws 1 blue then 1 green?

(A) $\frac{2}{9}$

(B) $\frac{15}{34}$

(C) $\frac{5}{9}$

(D) $\frac{7}{30}$

MM1D2b

55. If Sara has 6 shirts and 3 pairs of pants, how many different outfits does she have?

(A) 9

(B) 12

(C) 15

(D) 18

MM1D1b

56. Russell has a bag of black and red marbles. The probability that he first pulls out a black marble then pulls out a red marble is 34%. The probability he will pull out a black on the first try is 47%. What is the probability the second marble is red given the first marble is black?

(A) 53%

(B) 70%

(C) 72%

(D) 81%

MM1D2c

57. What is the probability that the spinner will stop on a shaded wedge or the number 1?

(A) $\frac{3}{7}$

(B) $\frac{5}{7}$

(C) $\frac{4}{7}$

(D) $\frac{6}{7}$

MM1D2d

58. Greg has test scores of 77, 90, 92, and 77 in English class. If he gets a 0 on his next paper, how will it affect the mean, median and mode of his scores?

(A) The mean will go down.
   The median will go down.
   The mode will remain the same.

(B) The mean will remain the same.
   The median will remain the same.
   The mode will remain the same.

(C) The mean, median and mode will all go down.

(D) The mean will go down.
   The mode and median will remain the same.

MM1D3a

59. What is the mean absolute deviation of the following set of numbers $\{3, 6, 0, 9, 1, 7, 2\}$?

(A) 2
(B) 2.5
(C) 2.86
(D) 5

MM1D4

60. $2\sqrt{x} + 1 = 7$. What is x?

(A) 3
(B) 4
(C) 9
(D) 16

MM1A3b

61. What are the zeros of $y = x^2 - 7x + 12$?

(A) $x = -4, 3$
(B) $x = 4, 3$
(C) $x = -4, -3$
(D) $x = 4, -3$

MM1A3a

62. A Republican politician running for mayor wants to take a poll to determine if he is the favored candidate in the upcoming election. The mayoral candidate is scheduled to give a speech not at all related to the election at an upcoming town celebration. For convenience sake, his staff polls all of the people in attendance at the speech as to who they are planning on voting for in the upcoming mayoral election. The staff reports to the candidate that, based on the recent poll, he is favored by voters. Are the results of this poll most likely accurate or are they most likely misleading?

(A) The results of the survey are accurate as long as at least 100 people were polled.

(B) The results of the poll are most likely misleading because the sample of people surveyed were voluntarily attending a speech given by the candidate and do not necessarily represent the population of the town.

(C) The results of the poll are most likely accurate because people are generally honest in declaring who they are going to vote for in elections.

(D) The results of the poll are misleading because recent advertisements are encouraging people to vote, creating a more opinionated population than usual.

MM1D3c

63. What is the solution to the equations $f(x) = x + 3$ and $g(x) = -2x$?

(A) $(3, 6)$
(B) $(1, 4)$
(C) $(-1, 2)$
(D) $(-3, 6)$

MM1A1i

64. $f(x) = x^2 - x - 2$ According to the graph, what are the solutions (zeros) to the equation?

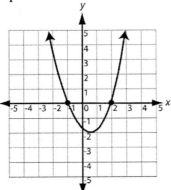

(A) $-2, 0$
(B) $-1, 1$
(C) $-1, 2$
(D) $-2, 1$

MM1A3c

65. What is the distance between $(0, 0)$ and $(3, 2)$?

(A) 13
(B) $\sqrt{13}$
(C) 4.12
(D) 3.555

MM1G1a

66. Solve for x: $x^2 - 3x - 10 = 0$

(A) $x = 5, 2$
(B) $x = 5, -2$
(C) $x = -5, 2$
(D) $x = -5, -2$

MM1A3d

67. What is the midpoint of $(-1, 2)$ and $(3, -6)$?

(A) $(1, 2)$
(B) $(2, 3)$
(C) $(3, 2)$
(D) $(1, -2)$

MM1G1c

68. Samantha is a student at Etowah High School. Look at the following histograms. The first histogram displays the number of sodas consumed each day by each student in Samantha's math class. The second histogram displays the number of sodas consumed each day by each of the students in the ninth-grade class at her school.

Find the difference in upper quartiles between the two data sets displayed in the histograms.

(A) 0

(B) 1

(C) 2

(D) 3

MM1D3b

69. Use the points $A = (1, -1)$, $B = (1, -5)$, $C = (4, -5)$ and the Pythagorean Theorem to find the distance between A and C. What is the distance?

(A) $\sqrt{5}$

(B) 2.3

(C) 5

(D) 25

MM1G1d

70. You are about to get a car for the first time. This means you will also get a licence plate. Each licence plate gets 3 numbers, then 3 letters. The numbers can be the digits from 0 to 9, and the letters can be any letter from A to Z. If there are no restrictions on what numbers or letters you can get, how many possible combinations of the licence plates are there?

(A) 108

(B) 17,576,000

(C) 760

(D) 12,245,000

MM1D1a

71. Graph the points $(2, 0)$, $(2, 2)$, $(0, 2)$, $(0, 0)$ to find out what shape they create. What shape do they create?

(A) a square

(B) a rectangle

(C) a triangle

(D) a rhombus

MM1G1e

72. Sean rolls a six-sided number cube with faces numbered 1 through 6, then tosses a coin. What is the probability that the coin has heads facing up, given that the number cube landed on 4?

(A) $\dfrac{1}{12}$

(B) $\dfrac{1}{3}$

(C) $\dfrac{1}{18}$

(D) $\dfrac{1}{2}$

MM1D2c

Evaluation Chart for the Diagnostic Mathematics Test

Directions: On the following chart, circle the question numbers that you answered incorrectly. Then turn to the appropriate topics (listed by chapters), read the explanations, and complete the exercises. Review the other chapters as needed. Finally, complete the *Passing the Georgia Mathematics I End-of-Course Test* Practice Tests to further review.

		Questions Part 1	Questions Part 2	Pages
Chapter 1:	Relations and Functions	2, 3, 4	38	15–31
Chapter 2:	Graphing Functions	1	37, 39	32–59
Chapter 3:	Roots		40	60–67
Chapter 4:	Polynomials	6	41, 42	68–89
Chapter 5:	Factoring	7, 8	43	90–112
Chapter 6:	Solving Quadratic Equations	25		113–116
Chapter 7:	Triangles	9, 11, 12, 13	44, 45, 46, 47, 49, 50	117–130
Chapter 8:	Quadrilaterals and Other Polygons	10	48	131–139
Chapter 9:	Logic and Geometric Proofs	14, 15	51, 52	140–147
Chapter 10:	Permutations and Combinations	17	55, 70	148–155
Chapter 11:	Probability	16, 18, 19, 20, 21, 36	53, 54, 56, 57, 72	156–175
Chapter 12:	Statistics	22, 23, 33, 34	58, 59, 62, 68	176–189
Chapter 13:	More Functions	5, 24	63	190–199
Chapter 14:	Solving Equations	26, 27, 29	60, 61, 64, 66	200–216
Chapter 15:	Coordinate Geometry	28, 30, 31, 32, 35	65, 67, 69, 71	217–229

NOTES

Chapter 1
Relations and Functions

This chapter covers the following Georgia Performance Standards:

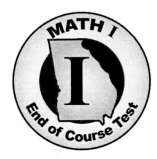

MM1A	Algebra	MM1A1a
		MM1A1d
		MM1A1e
		MM1A1f

1.1   Relations

A **relation** is a set of ordered pairs. The set of the first members of each ordered pair is called the **domain** of the relation. The set of the second members of each ordered pair is called the **range**.

Example 1:   State the domain and range of the following relation:

$$\{(2, 4), (3, 7), (4, 9), (6, 11)\}$$

Solution: Domain: $\{2, 3, 4, 6\}$ the first member of each ordered pair

Range: $\{4, 7, 9, 11\}$ the second member of each ordered pair

State the domain and range for each relation.

1. $\{(2, 5), (9, 12), (3, 8), (6, 7)\}$

2. $\{(12, 4), (3, 4), (7, 12), (26, 19)\}$

3. $\{(4, 3), (7, 14), (16, 34), (5, 11)\}$

4. $\{(2, 45), (33, 43), (98, 9), (43, 61), (67, 54)\}$

5. $\{(78, 14), (29, 67), (84, 49), (16, 18), (98, 46)\}$

6. $\{(-8, 16), (23, -7), (-4, -9), (16, -8), (-3, 6)\}$

7. $\{(-7, -4), (-3, 16), (-4, 17), (-6, -8), (-8, 12)\}$

8. $\{(-1, -2), (3, 6), (-7, 14), (-2, 8), (-6, 2)\}$

9. $\{(0, 9), (-8, 5), (3, 12), (-8, -3), (7, 18)\}$

10. $\{(58, 14), (44, 97), (74, 32), (6, 18), (63, 44)\}$

When given an equation in two variables, the domain is the set of x values that satisfies the equation. The range is the set of y values that satisfies the equation.

Example 2:   Find the range of the relation $3x = y + 2$ for the domain $\{-1, 0, 1, 2, 3\}$.
Solve the equation for each value of x given. The result, the y values, will be the range.

<table>
<tr><th colspan="2">Given:</th><th colspan="2">Solution:</th></tr>
<tr><td>x</td><td>y</td><td>x</td><td>y</td></tr>
<tr><td>-1</td><td></td><td>-1</td><td>-5</td></tr>
<tr><td>0</td><td></td><td>0</td><td>-2</td></tr>
<tr><td>1</td><td></td><td>1</td><td>1</td></tr>
<tr><td>2</td><td></td><td>2</td><td>4</td></tr>
<tr><td>3</td><td></td><td>3</td><td>7</td></tr>
</table>

The range is $\{-5, -2, 1, 4, 7\}$.

Find the range of each relation for the given domain.

	Relation	**Domain**	**Range**
1.	$y = 5x$	$\{1, 2, 3, 4\}$	
2.	$y = |x|$	$\{-3, -2, -1, 0, 1\}$	
3.	$y = 3x + 2$	$\{0, 1, 3, 4\}$	
4.	$y = -|x|$	$\{-2, -1, 0, 1, 2\}$	
5.	$y = -2x + 1$	$\{0, 1, 3, 4\}$	
6.	$y = 10x - 2$	$\{-2, -1, 0, 1, 2\}$	
7.	$y = 3|x| + 1$	$\{-2, -1, 0, 1, 2\}$	
8.	$y - x = 0$	$\{1, 2, 3, 4\}$	
9.	$y - 2x = 0$	$\{1, 2, 3, 4\}$	
10.	$y = 3x - 1$	$\{0, 1, 3, 4\}$	
11.	$y = 4x + 2$	$\{0, 1, 3, 4\}$	
12.	$y = 2|x| - 1$	$\{-2, -1, 0, 1, 2\}$	

1.2 Functions

Some relations are also **functions**. A relation is a function if **for every element in the domain, there is exactly one element in the range**. In other words, for each value for x there is only one unique value for y.

Example 3: $\{(2,4), (2,5), (3,4)\}$ is **NOT** a function because in the first pair, 2 is paired with 4, and in the second pair, 2 is paired with 5. The 2 can be paired with only one number to be a function. In this example, the x value of 2 has more than one value for y: 4 and 5.

Example 4: $\{(1,2), (3,2), (5,6)\}$ **IS** a function. Each first number is paired with only one second number. The 2 is repeated as a second number, but the relation remains a function.

Determine whether the ordered pairs of numbers below represent a function. Write "F" if it is a function. Write "NF" if it is not a function.

1. $\{(-1,1), (-3,3), (0,0), (2,2)\}$ _____

2. $\{(-4,-3), (-2,-3), (-1,-3), (2,-3)\}$ _____

3. $\{(5,-1), (2,0), (2,2), (5,3)\}$ _____

4. $\{(-3,3), (0,2), (1,1), (2,0)\}$ _____

5. $\{(-2,-5), (-2,-1), (-2,1), (-2,3)\}$ _____

6. $\{(0,2), (1,1), (2,2), (4,3)\}$ _____

7. $\{(4,2), (3,3), (2,2), (0,3)\}$ _____

8. $\{(-1,-1), (-2,-2), (3,-1), (3,2)\}$ _____

9. $\{(2,-2), (0,-2), (-2,0), (1,-3)\}$ _____

10. $\{(2,1), (3,2), (4,3), (5,-1)\}$ _____

11. $\{(-1,0), (2,1), (2,4), (-2,2)\}$ _____

12. $\{(1,4), (2,3), (0,2), (0,4)\}$ _____

13. $\{(0,0), (1,0), (2,0), (3,0)\}$ _____

14. $\{(-5,-1), (-3,-2), (-4,-9), (-7,-3)\}$ _____

15. $\{(8,-3), (-4,4), (8,0), (6,2)\}$ _____

16. $\{(7,-1), (4,3), (8,2), (2,8)\}$ _____

17. $\{(4,-3), (2,0), (5,3), (4,1)\}$ _____

18. $\{(2,-6), (7,3), (-3,4), (2,-3)\}$ _____

19. $\{(1,1), (3,-2), (4,16), (1,-5)\}$ _____

20. $\{(5,7), (3,8), (5,3), (6,9)\}$ _____

1.3   Function Notation

Function notation is used to represent relations which are functions. Some commonly used letters to represent functions include f, g, h, F, G, and H.

Example 5: $f(x) = 2x - 1$; find $f(-3)$

Step 1:   Find $f(-3)$ means to replace x with -3 in the relation $2x - 1$.
$f(-3) = 2(-3) - 1$

Step 2:   Solve $f(-3)$. $f(-3) = 2(-3) - 1 = -6 - 1 = -7$
$f(-3) = -7$

Example 6: $g(x) = 4 - 2x^2$; find $g(2)$

Step 1:   Replace x with 2 in the relation $4 - 2x^2$.
$g(2) = 4 - 2(2)^2$

Step 2:   Solve $g(2)$. $g(2) = 4 - 2(2)^2 = 4 - 2(4) = 4 - 8 = -4$
$g(2) = -4$

Find the solutions for each of the following.

1. $F(x) = 2 + 3x^2$; find $F(3)$

2. $f(x) = 4x + 6$; find $f(-4)$

3. $H(x) = 6 - 2x^2$; find $H(-1)$

4. $g(x) = -3x + 7$; find $g(-3)$

5. $f(x) = -5 + 4x$; find $f(7)$

6. $G(x) = 4x^2 + 4$; find $G(0)$

7. $f(x) = 7 - 6x$; find $f(-4)$

8. $h(x) = 2x^2 + 10$; find $h(5)$

9. $F(x) = 7 - 5x$; find $F(2)$

10. $f(x) = -4x^2 + 5$; find $f(-2)$

1.4   Function Tables

Functions can also use a variable such as n to be the input of the function and $f(n)$, read "f of n," to represent the output of the function.

Example 7:   Function rule: $3n + 4$

n	$f(n)$
-1	1
0	4
1	7
2	10

Fill in the tables for each function rule below.

1. rule: $2(n-5)$

n	$f(n)$
0	
1	
2	
3	

4. rule: $2x(x-1)$

x	$f(x)$
1	
2	
3	
4	

7. rule: $n(n+2)$

n	$f(n)$
1	
2	
3	
4	

2. rule: $3x(x-4)$

x	$f(x)$
0	
1	
2	
3	

5. rule: $\dfrac{1}{n+3}$

n	$f(n)$
1	
2	
3	
4	

8. rule: $2x-3$

x	$f(x)$
1	
2	
3	
4	

3. rule: $\dfrac{2-n}{2}$

n	$f(n)$
0	
2	
4	
6	
8	

6. rule: $4x-x$

x	$f(x)$
-2	
-1	
0	
1	
2	

9. rule: $3-2n$

n	$f(n)$
-2	
-1	
0	
1	
2	

1.5   Recognizing Functions

Recall that a relation is a function with only one y value for every x value. We can depict functions in many ways including through graphs.

Example 8:

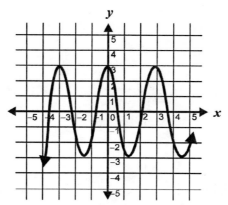

This graph **IS** a function because it has only one y value for each value of x.

Example 9:

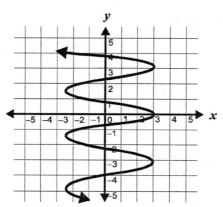

This graph is **NOT** a function because there is more than one y value for each value of x.

Hint: An easy way to determine a function from a graph is to do a vertical line test. First, draw a vertical line that crosses over the whole graph. If the line crosses the graph more than one time, then it is not a function. If it only crosses it once, it is a function. Take Example 9 above:

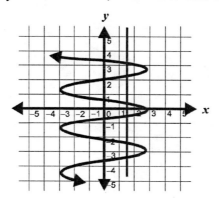

Since the vertical line passes over the graph six times, it is not a function.

Determine whether or not each of the following graphs is a function. If it is, write function on the line provided. If it is not a function, write NOT a function on the line provided.

1.

_____

4.

_____

2.

_____

5.

_____

3.

_____

6.

_____

7.

_____

8.

_____

9.

_____

10.

_____

11.

_____

12.

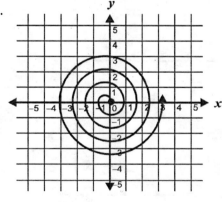

_____

1.6   Characteristics of Functions

There are many different types of functions. For our purposes, however, we will only look at the six basic functions and their characteristics.

The Linear Function: $f(x) = x$

Standard Form: $y = mx + b$, where m is the constant slope and b is the y-intercept

Characteristics: straight line

Domain: All Real Numbers

Range: All Real Numbers

Example 10: $f(x) = 2x + 7$; slope = 2, y-intercept = 7

x	−2	−1	0	1	2
$f(x)$	3	5	7	9	11

The Quadratic Function: $f(x) = x^2$

Standard Form: $f(x) = ax^2 + bx + c$, where $a \neq 0$, a, b, and c are real numbers, and c is the y-intercept

Characteristics: Has a vertex, parabola (U) shape

Domain: All Real Numbers

Range: $(0, \infty)$

Example 11: $f(x) = x^2 + 3x + 2$

x	−2	−1	0	1	2
$f(x)$	0	0	2	6	12

The Cubic Function: $f(x) = x^3$

Standard Form: $f(x) = ax^3 + bx^2 + cx + d$, where $a \neq 0$

Characteristics: N shape

Domain: All Real Numbers

Range: All Real Numbers

Example 12: $f(x) = x^3 + 3x^2 + 6x + 2$

x	−2	−1	0	1	2
$f(x)$	−6	−2	2	12	34

The Square Root Function: $f(x) = \sqrt{x}$

Characteristics: Never negative (neither x nor answer is negative), always contains a radical

Domain: $(0, \infty)$

Range: $(0, \infty)$

Example 13: $f(x) = \sqrt{x} + 2$

x	0	1	2	3	4
$f(x)$	2	3	$2 + \sqrt{2}$	$2 + \sqrt{3}$	4

The Absolute Value Function: $f(x) = |x|$

Characteristics: Answer usually positive unless outside value is being subtracted and is larger than inside value (i.e. $|-2| - 3 = -1$), V shape

Domain: All Real Numbers

Range: $(0, \infty)$

Example 14: $f(x) = |x| + 2$

x	-2	-1	0	1	2
$f(x)$	4	3	2	3	4

$f(x) = \dfrac{1}{x}$

Characteristics: Vertical asymptote at $x = 0$, undefined at $x = 0$

Domain: $(-\infty, 0) \cup (0, \infty)$

Range: $(-\infty, 0) \cup (0, \infty)$

Example 15: $f(x) = \dfrac{1}{x} + 2$

x	-2	-1	0	1	2
$f(x)$	$1\frac{1}{2}$	1	Undefined	3	$2\frac{1}{2}$

Determine the function described by the characteristics.

1. Its shape is a parabola.

2. It NEVER includes negative numbers.

3. It has an N shape.

4. It is undefined at $x = 0$.

5. The values of y are always positive, but x can be negative.

6. It is a straight line.

7. Of the 6 functions, how many have domains of all real numbers?

8. How many have limited ranges (not all real numbers)?

1.7 Sequences

In each of the examples below, there is a sequence of numbers that follows a pattern. Think of the sequence of numbers like the output for a function. You must find the pattern (or function) that holds true for each number in the sequence. Once you determine the pattern, you can find the next number in the sequence or any number in the sequence.

	Sequence	Pattern	Next Number	20th number in the sequence
Example 16:	$3, 4, 5, 6, 7$	$n+2$	8	22

In number patterns, the sequence is the output (the range). The input (the domain) can be the set of whole numbers starting with 1. But, you must determine the "rule" or pattern. Look at the table below.

input	sequence
1	3
2	4
3	5
4	6
5	7

What pattern or "rule" can you come up with that gives you the first number in the sequence, 3, when you input 1? $n+2$ will work because when $n=1$, the first number in the sequence $=3$. Does this pattern hold true for the rest of the numbers in the sequence? Yes, it does. When $n=2$, the second number in the sequence $=4$. When $n=3$, the third number in the sequence $=5$, and so on. Therefore, $n+2$ is the pattern. Even without knowing the algebraic form of the pattern, you could figure out that 8 is the next pattern in the sequence. To find the 20th number in the pattern, use $n=20$ to get 22.

	Sequence	Pattern	Next Number	20th number in the sequence
Example 17:	$1, 4, 9, 16, 25$	n^2	36	400
Example 18:	$-2, -4, -6, -8, -10$	$-2n$	-12	-40

Find the pattern and the next number in each of the sequences below.

	Sequence	Pattern	Next Number	20th number in the sequence
1.	$-2, -1, 0, 1, 2$			
2.	$5, 6, 7, 8, 9$			
3.	$3, 7, 11, 15, 19$			
4.	$-3, -6, -9, -12, -15$			
5.	$3, 5, 7, 9, 11$			
6.	$2, 4, 8, 16, 32$			
7.	$1, 8, 27, 64, 125$			
8.	$0, -1, -2, -3, -4$			
9.	$2, 5, 10, 17, 26$			
10.	$4, 6, 8, 10, 12$			

1.8   Using the $\sum$ Symbol

The Σ is used to write the sum of a series.

Series A $1 + 2 + 3 + \ldots + n + \ldots$

Series A starts with 1, +2, +3. Obviously the next item in the series would be 4.

Series B $1 + 4 + 7 + \ldots + (3n - 2) + \ldots$

Series B starts with 1, +4, +7 . . . $(3n - 2)$. The expression in parenthesis shows how the pattern continues. The "n" in the expression is the item we're trying to find. We have 3 items: 1, 4, and 7. We are looking for the 4th item in the series so "n" = 4. $3n - 2$ where "n" = 4 is $3(4) - 2 = 10$. The next item in the series is 10.

Series B continues $1 + 4 + 7 + 10 + \ldots + (3n - 2) + \ldots$ to ∞ (infinity).

Series C $1 + 4 + 9 + \ldots + (n^2) + \ldots$

Replacing "n" with 4 in series C to find the 4th item in the series we have $4^2 = 16$. So now the series is $1 + 4 + 9 + 16 + \ldots + (n^2) + \ldots$

The sum of the first "n" terms of a series is called the **partial sum**, S_n. For example, in the series $3 + 6 + 9 + \ldots + 3n \ldots$

$$S_1 = 3 \qquad S_2 = 3 + 6 = 9 \qquad S_3 = 3 + 6 + 9 = 18$$

The Greek letter Σ, sigma, can be used to represent a sum. For example the third partial sum, S_3 of the series above can also be written as:

$$\sum_{k=1}^{3} 3k$$

So the series starts at $k = 1$ and ends at $k = 3$ and formula for the pattern is $3k$.

First $k = 1$ then 2, then 3. In expanded form we have: $3 + 6 + 9 = 18$. (We found the partial sum of the series.)

For each of the following write the Σ notation for each statement.

1. The sum of the series with the general term $k + 2$ from $k = 1$ to $k = 8$.

2. The sum of the series with the general term k^2 from $k = 1$ to $k = 5$.

3. The sum of the series with the general term $2k + 3$ from $k = 1$ to $k = 10$.

4. The sum of the series with the general term $2k - 6$ from $k = 3$ to $k = 8$.

5. The sum of the series with the general term $2m - 6$ from $m = 0$ to $m = 4$.

6. The sum of the series with the general term $2g^2$ from $g = 2$ to $g = 5$.

7. The sum of the series with the general term $g^2 + 2$ from $g = 3$ to $g = 7$.

8. The sum of the series with the general term $3h - 4$ from $h = 0$ to $h = 6$.

Example 19:   Write $\displaystyle\sum_{n=2}^{5} 3n$ in expanded form, then solve.

Step 1:   In this problem, $n = 2, 3, 4,$ and 5. Plug the values for n into the expression $3n$. We have $3 \cdot 2 + 3 \cdot 3 + 3 \cdot 4 + 3 \cdot 5$.

Step 2:   Simplify the expanded form and solve.
$3 \cdot 2 + 3 \cdot 3 + 3 \cdot 4 + 3 \cdot 5 = 6 + 9 + 12 + 15 = 42.$

Write each of the following in expanded form and find the sums.

1. $\displaystyle\sum_{n=3}^{5} 2n$

2. $\displaystyle\sum_{n=0}^{4} n^2$

3. $\displaystyle\sum_{n=0}^{4} 2n$

4. $\displaystyle\sum_{n=1}^{5} (2n - 1)$

5. $\displaystyle\sum_{n=0}^{3} (3n + 1)$

6. $\displaystyle\sum_{n=3}^{6} (3^n + 1)$

7. $\displaystyle\sum_{n=0}^{5} (2n + 1)$

8. $\displaystyle\sum_{n=1}^{4} (n - 6)$

9. $\displaystyle\sum_{n=4}^{6} (3n - 2)$

10. $\displaystyle\sum_{n=2}^{5} (6n - 4)$

Chapter 1 Review

1. What is the domain of the following relation? $\{(-1, 2), (2, 5), (4, 9), (6, 11)\}$

2. What is the range of the following relation? $\{(0, -2), (-1, -4), (-2, 6), (-3, -8)\}$

3. Find the range of the relation $y = 5x$ for the domain $\{0, 1, 2, 3, 4\}$.

4. Find the range of the relation $y = \dfrac{3(x-2)}{5}$ for the domain $\{-8, -3, 7, 12, 17\}$.

5. Find the range of the relation $y = 10 - 2x$ for the domain $\{-8, -4, 0, 4, 8\}$.

6. Find the range of the relation $y = \dfrac{4 + x}{3}$ for the domain $\{-7, -1, 2, 5, 8\}$.

For each of the following relations given in questions 7–11, write F if it is a function and NF if it is not a function.

7. $\{(1, 2), (2, 2), (3, 2)\}$

8. $\{(-1, 0), (0, 1), (1, 2), (2, 3)\}$

9. $\{(2, 1), (2, 2), (2, 3)\}$

10. $\{(1, 7), (2, 5), (3, 6), (2, 4)\}$

11. $\{(0, -1), (-1, -2), (-2, -3), (-3, -4)\}$

For questions 12–17, find the range of the following functions for the given value of the domain.

12. For $g(x) = 2x^2 - 4x$; find $g(-1)$

13. For $h(x) = 3x(x-4)$; find $h(3)$

14. For $f(n) = \dfrac{1}{n+3}$; find $f(4)$

15. For $G(n) = \dfrac{2-n}{2}$; find $G(8)$

16. For $H(x) = 2x(x-1)$; find $H(4)$

17. For $f(x) = 7x^2 + 3x - 2$; find $f(2)$

Fill in the following function tables.

18. Rule: $\dfrac{(4-2n)}{2}$

n	$f(n)$
0	
1	
2	
3	
4	

19. Rule: $2n(n+1)$

n	$f(n)$
0	
1	
2	
3	
4	

20. Rule: $6n-3$

n	$f(n)$
2	
3	
4	
5	
6	

Find the pattern for the following number sequences, and then find the nth number requested.

21. 0, 1, 2, 3, 4   pattern_____

22. 0, 1, 2, 3, 4   20th number_____

23. 1, 3, 5, 7, 9   pattern_____

24. 1, 3, 5, 7, 9   25th number_____

25. 3, 6, 9, 12, 15   pattern_____

26. 3, 6, 9, 12, 15   30th number_____

Answer the following questions about the six basic functions.

27. What is the domain of a linear function?

28. What is the range of $f(x) = \dfrac{1}{x}$?

29. What function has a constant rate of change?

30. What is the standard form of a quadratic function?

31. What is the range of the square root function?

Chapter 1 Test

1. Which of the following graphs is a function?

(A)

(B)

(C)

(D)
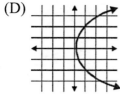

2. The function rule is $3x(x+5)$.

x	$f(x)$
-2	?

(A) -18
(B) 18
(C) -42
(D) 42

3. What is the range of the following relation?

$\{(1,2)(4,9)(7,8)(10,13)\}$

(A) $\{1, 4, 7, 10\}$
(B) $\{2, 9, 8, 13\}$
(C) $\{3, 13, 15, 23\}$
(D) $\{1, 3, 1, 3\}$

4. Which function is not linear?

(A) $f(x) = -3x - 7$
(B) $f(x) = 8$
(C) $f(x) = \frac{2}{3}x - 5$
(D) $f(x) = 2x^2$

5. Which function does not have a zero?

(A) $f(x) = x - 3$
(B) $f(x) = -7$
(C) $f(x) = 2x + 3$
(D) $f(x) = -3x + 2$

6. Which of the following relations is a function?

(A) $\{(0,-1)(1,-1)(0,-2)\}$
(B) $\{(-1,1)(-1,-1)(0,0)\}$
(C) $\{(2,1)(1,0)(0,-1)\}$
(D) $\{(-1,1)(-1,0)(-1,-1)\}$

7. Find the range of the following function for the domain $\{-2, -1, 0, 3\}$.

$$y = \frac{2+x}{4}$$

(A) $\left\{0, \frac{3}{4}, 1, \frac{5}{4}\right\}$

(B) $\left\{0, \frac{1}{4}, \frac{1}{2}, \frac{5}{4}\right\}$

(C) $\left\{1, -\frac{1}{4}, \frac{1}{2}, \frac{5}{4}\right\}$

(D) $\left\{\frac{1}{4}, \frac{3}{4}, \frac{1}{2}, \frac{5}{4}\right\}$

8. For $f(x) = 3x^2 - 5x$, find $f(-3)$.

(A) 12
(B) -6
(C) 42
(D) 3

9. What is the next number in this sequence?

    0.03, 0.12, 0.48, 1.92, _____

    (A) 1.95
    (B) 3.36
    (C) 5.08
    (D) 7.68

10. What is the next number in this sequence?

    2, 16, 128, 1024, _____

    (A) 1, 920
    (B) 8, 192
    (C) 11, 586
    (D) 16, 384

11. Which function has only one vertex?

    (A) the linear function
    (B) the quadratic function
    (C) the cubic function
    (D) the square root function

12. A radical is always in what kind of function?

    (A) the linear function
    (B) the absolute value function
    (C) the square root function
    (D) $\dfrac{1}{x}$

13. $f(x) = ax^3 + bx^2 + cx + d$ is the standard form of which function?

    (A) the cubic function
    (B) the square root function
    (C) the quadratic function
    (D) the linear function

14. Which function has a V shape?

    (A) the linear function
    (B) the absolute value function
    (C) the cubic function
    (D) the square root function

15. Which function is undefined at $x = 0$?

    (A) the cubic function
    (B) the square root function
    (C) the absolute value function
    (D) $\dfrac{1}{x}$

16. Which equation is a function of x?

    (A) $f(b) = 3b + 2$
    (B) $f(x) = 3x + 2$
    (C) $f(a) = 3a + 2$
    (D) $f(z) = 3z + 2$

17. Which is the equation $y = 7x^2 + 2x + 3$ as a function of m?

    (A) $m = 7x^2 + 2x + 3$
    (B) $y = 7m^2 + 2m + 3$
    (C) $m(x) = 7m^2 + 2m + 3$
    (D) $f(m) = 7m^2 + 2m + 3$

Chapter 2
Graphing Functions

This chapter covers the following Georgia Performance Standards:

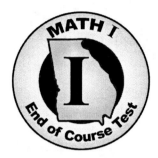

MM1A	Algebra	MM1A1b
		MM1A1c
		MM1A1d
		MM1A1e
		MM1A1g

2.1 Graphing Linear Equations

In addition to graphing ordered pairs, the Cartesian plane can be used to graph the solution set for an equation. Any equation with two variables that are both to the first power is called a **linear equation.** The graph of a linear equation will always be a straight line.

Example 1: Graph the solution set for $x + y = 7$.

Step 1: Make a list of some pairs of numbers that will work in the equation.

$$\begin{aligned} &x + y = 7 \\ &\overline{4 + 3 = 7} \qquad (4, 3) \\ &-1 + 8 = 7 \qquad (-1, 8) \\ &5 + 2 = 7 \qquad (5, 2) \\ &0 + 7 = 7 \qquad (0, 7) \end{aligned} \Bigg\} \text{ordered pair solutions}$$

Step 2: Plot these points on a Cartesian plane.

Step 3: By passing a line through these points, we graph the solution set for $x + y = 7$. This means that every point on the line is a solution to the equation $x + y = 7$. For example, $(1, 6)$ is a solution, so the line passes through the point $(1, 6)$.

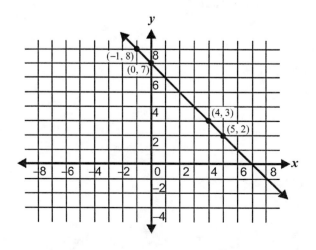

Make a table of solutions for each linear equation below. Then plot the ordered pair solutions on graph paper. Draw a line through the points. (If one of the points does not line up, you have made a mistake.)

1. $x + y = 6$

2. $y = x + 1$

3. $y = x - 2$

4. $x + 2 = y$

5. $x - 5 = y$

6. $x - y = 0$

Example 2: Graph the equation $y = 2x - 5$.

Step 1: This equation has 2 variables, both to the first power, so we know the graph will be a straight line. Substitute some numbers for x or y to find pairs of numbers that satisfy the equation. For the above equation, it will be easier to substitute values of x in order to find the corresponding value for y. Record the values for x and y in a table.

	x	y
If x is 0, y would be -5	0	-5
If x is 1, y would be -3	1	-3
If x is 2, y would be -1	2	-1
If x is 3, y would be 1	3	1

Step 2: Graph the ordered pairs, and draw a line through the points.

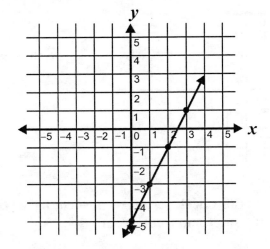

Find pairs of numbers that satisfy the equations below, and graph the line on graph paper.

1. $y = -2x + 2$

2. $2x - 2 = y$

3. $-x + 3 = y$

4. $y = x + 1$

5. $4x - 2 = y$

6. $y = 3x - 3$

7. $x = 4y - 3$

8. $2x = 3y + 1$

9. $x + 2y = 4$

2.2   Slope-Intercept Form of a Line

An equation that contains two variables, each to the first degree, is a **linear equation**. The graph for a linear equation is a straight line. To put a linear equation in slope-intercept form, solve the equation for y. This form of the equation shows the slope and the y-intercept. Slope-intercept form follows the pattern of $y = mx + b$. The "m" represents slope, and the "b" represents the y-intercept. The y-intercept is the point at which the line crosses the y-axis. The constant is called the **slope** of the line.

Example 3:   Put the equation $2x + 3y = 15$ in slope-intercept form. What is the slope of the line? What is the y-intercept? Graph the line.

Step 1:   Solve for y:

$$
\begin{array}{rcr}
2x + 3y &=& 15 \\
-2x & & -2x \\
\hline
\dfrac{3y}{3} &=& -\dfrac{2x}{3} + \dfrac{15}{3}
\end{array}
$$

slope-intercept form: $y = -\frac{2}{3}x + 5$

The slope is $-\frac{2}{3}$ and the y-intercept is 5.

Step 2:   Knowing the slope and the y-intercept, we can graph the line.

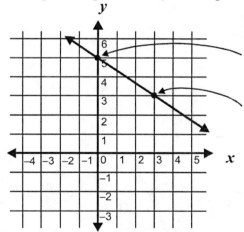

The y-intercept is 5, so the line passes through the point $(0, 5)$ on the y-axis.

The slope is $-\frac{2}{3}$, so go down 2 and over 3 to get a second point.

Put each of the following equations in slope-intercept form by solving for y. On your graph paper, graph the line using the slope and y-intercept.

1. $4x - y = 5$
2. $2x + 4y = 16$
3. $3x - 2y = 10$
4. $x + 3y = -12$
5. $6x + 2y = 0$

6. $8x - 5y = 10$
7. $-2x + y = 4$
8. $-4x + 3y = 12$
9. $-6x + 2y = 12$
10. $x - 5y = 5$

11. $3x - 2y = -6$
12. $3x + 4y = 2$
13. $-x = 2 + 4y$
14. $2x = 4y - 2$
15. $6x - 3y = 9$

16. $4x + 2y = 8$
17. $6x - y = 4$
18. $-2x - 4y = 8$
19. $5x + 4y = 16$
20. $6 = 2y - 3x$

2.3 Graphing Quadratic Equations

Equations that you may encounter on the GA Math I EOCT will possibly involve variables which are squared (raised to the second power). The best way to find values for the x and y variables in an equation is to plug one number into x, and then find the corresponding value for y just as you did at the beginning of this chapter. Then, plot the points and draw a line through the points.

Example 4: Graph $y = x^2$.

Step 1: Make a table and find several values for x and y.

x	y
-2	4
-1	1
0	0
1	2
2	4

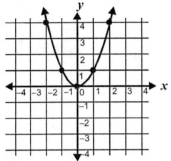

Step 2: Plot the points, and draw a curve through the points. Notice the shape of the curve. This type of curve is called a **parabola**. Equations with one squared term will be parabolas.

Example 5: Graph the equation $y = -2x^2 + 4$.

Step 1: Make a table and find several values for x and y.

x	y
-2	-4
-1	2
0	4
1	2
2	-4

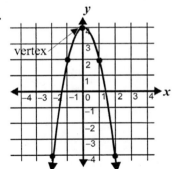

Step 2: Plot the points, and draw a curve through the points.

Note: In the equation $y = ax^2 + c$, changing the value of a will widen or narrow the parabola around the y-axis. If the value of a is a negative number, the parabola will be reflected across the x-axis (the vertex will be at the top of the parabola instead of at the bottom.) If $a = 0$, the graph will be a straight line, not a parabola. Changing the value of c will move the vertex of the parabola from the origin to a different point on the y-axis.

Graph the equations below on a Cartesian plane.

1. $y = 2x^2$
2. $y = 3 - x^2$
3. $y = x^2 - 2$
4. $y = -2x^2$
5. $y = x^2 + 3$
6. $y = -3x^2 + 2$
7. $y = 3x^2 - 5$
8. $y = x^2 + 1$
9. $y = -x^2 - 6$
10. $y = -x^2$
11. $y = 2x^2 - 1$
12. $y = 2 - 2x^2$

2.4   Graphing Basic Functions

A graph is an image that shows the relationship between two or more variables. In this section, we will learn how to graph six basic functions.

Example 6: $f(x) = x$

The notation $f(x)$ is the same as y, it just means f as a function of x

The easiest way to begin graphing this is to draw a table of values. To create an accurate graph, you should choose at least 5 values for your table. These values are now your (x, y) ordered pair and they are ready to be plotted.

x	$f(x)$
-2	-2
-1	-1
0	0
1	1
2	2

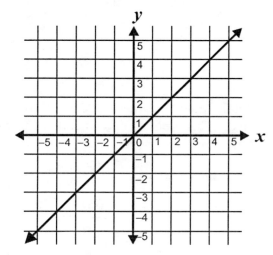

This is called the graph of a **linear function**.

Example 7: $f(x) = x^2$

Exponential functions can be graphed the same way as linear functions.

x	$f(x)$
-2	4
-1	1
0	0
1	1
2	4

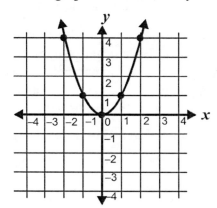

This is called the graph of a **quadratic function**.

Example 8: $f(x) = x^3$

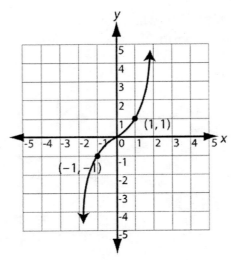

x	$f(x)$
-2	-8
-1	-1
0	0
1	1
2	8

This is called the graph of a **cubic function**.

Example 9: $f(x) = \sqrt{x}$

An equation like this will be easier to graph if you choose values of x that are perfect squares and because you can't take the square root of a negative, there is no need to select negative values.

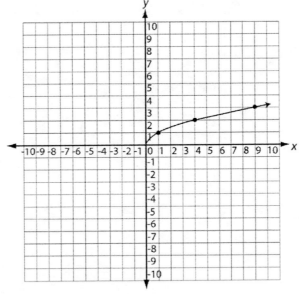

x	$f(x)$
0	0
1	1
4	2
9	3

This is called the graph of a **square root function**.

Example 10: $f(x) = |x|$

Don't forget, when you take the absolute value of a negative number, it becomes positive!

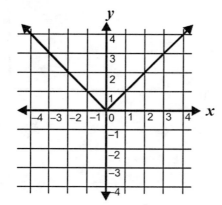

x	$f(x)$
-2	2
-1	1
0	0
1	1
2	2

This is called the graph of an **absolute value function**.

Example 11: $f(x) = \dfrac{1}{x}$

Don't forget, anything over zero is undefined and therefore does not exist!

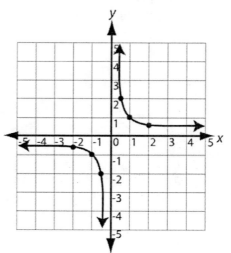

x	$f(x)$
-2	$-\frac{1}{2}$
-1	-1
$-\frac{1}{2}$	-2
0	Does not exist
$\frac{1}{2}$	2
1	1
2	$\frac{1}{2}$

Graph the following.

1. $f(x) = 2x$

2. $f(x) = 3x^3$

3. $f(x) = -\left|\frac{1}{2}x\right|$

4. $f(x) = |4x|$

5. $f(x) = \sqrt{2x}$

6. $f(x) = \dfrac{6}{x}$

7. $f(x) = \frac{1}{3}x^2$

8. $f(x) = -4x^2$

9. $f(x) = -\dfrac{2}{x}$

10. $f(x) = -\sqrt{x}$

2.5   Understanding Slope

The slope of a line refers to how steep a line is. Slope is also defined as the rate of change. When we graph a line using ordered pairs, we can easily determine the slope. Slope is often represented by the letter m.

$$\text{The formula for slope of a line is: } m = \frac{y_2 - y_1}{x_2 - x_1} \text{ or } \frac{\text{rise}}{\text{run}}$$

Example 12:   What is the slope of the following line that passes through the ordered pairs $(-4, -3)$ and $(1, 3)$?

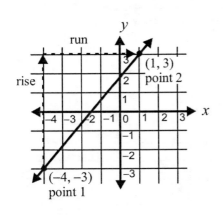

y_2 is 3, the y-coordinate of point 2.

y_1 is -3, the y-coordinate of point 1.

x_2 is 1, the x-coordinate of point 2.

x_1 is -4, the x-coordinate of point 1.

Use the formula for slope given above:

$$m = \frac{3 - (-3)}{1 - (-4)} = \frac{6}{5}$$

The slope is $\frac{6}{5}$. This shows us that we can go up 6 (rise) and over 5 to the right (run) to find another point on the line.

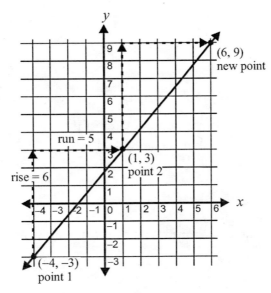

Example 13:   Find the slope of a line through the points $(-2, 3)$ and $(1, -2)$. It doesn't matter which pair we choose for point 1 and point 2. The answer is the same.

Let point 1 be $(-2, 3)$
Let point 2 be $(1, -2)$

$$\text{slope} = \frac{(y_2 - y_1)}{(x_2 - x_1)} = \frac{-2 - 3}{1 - (-2)} = \frac{-5}{3}$$

When the slope is negative, the line will slant up to the left. For this example, the line will go **down** 5 units and then over 3 units to the **right**.

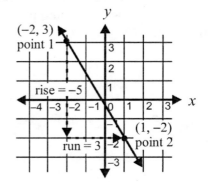

Example 14:   What is the slope of a line that passes through $(1, 1)$ and $(3, 1)$?

$$\text{slope} = \frac{1 - 1}{3 - 1} = \frac{0}{2} = 0$$

When $y_2 - y_1 = 0$, the slope will equal 0, and the line will be horizontal.

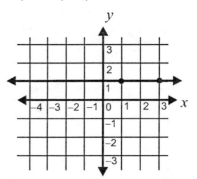

Example 15:   What is the slope of a line that passes through $(2, 1)$ and $(2, -3)$?

$$\text{slope} = \frac{-3 - 1}{2 - 2} = \frac{-4}{0} = \text{undefined}$$

When $x_2 - x_1 = 0$, the denominator of the slope is 0, then the slope is undefined, and the line will be vertical.

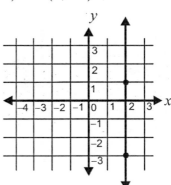

The following lines summarize what we know about slope.

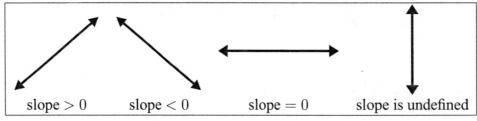

slope > 0 slope < 0 slope $= 0$ slope is undefined

Find the slope of the line that goes through the following pairs of points. Then, using graph paper, graph the line through the two points, and label the rise and run. (See Examples 12–15).

1. $(2, 3)$ $(4, 5)$

2. $(1, 3)$ $(2, 5)$

3. $(-1, 2)$ $(4, 1)$

4. $(1, -2)$ $(4, -2)$

5. $(3, 0)$ $(3, 4)$

6. $(3, 2)$ $(-1, 8)$

7. $(4, 3)$ $(2, 4)$

8. $(2, 2)$ $(1, 5)$

9. $(3, 4)$ $(1, 2)$

10. $(3, 2)$ $(3, 6)$

11. $(6, -2)$ $(3, -2)$

12. $(1, 2)$ $(3, 4)$

13. $(-2, 1)$ $(-4, 3)$

14. $(5, 2)$ $(4, -1)$

15. $(1, -3)$ $(-2, 4)$

16. $(2, -1)$ $(3, 5)$

2.6   Comparing Rates of Change

The rate of change is the value at which the quantity, y, is changing in respect to quantity, x.

$$\text{Rate of Change} = \text{change in } y \div \text{change in } x \left(\frac{y_2 - y_1}{x_2 - x_1} \right)$$

In a linear function, the rate of change, also known as the slope, remains constant. This is great for describing an object moving at a constant speed, but that is not always the case. That is why, in quadratic functions, the rate of change varies. Quadratic equations have a variable rate of change.

Example 16:   Linear Functions

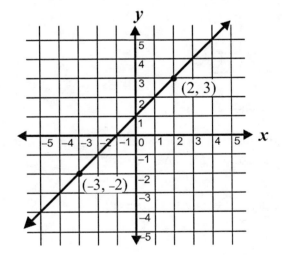

Find the slope: $\dfrac{y_2 - y_1}{x_2 - x_1} = \dfrac{3 - (-2)}{2 - (-3)} = \dfrac{5}{5} = 1$

This value shows that every time x increases by one, y does the same. The slope of a linear equation is always constant.

Example 17:   Quadratic Function, $y = x^2$

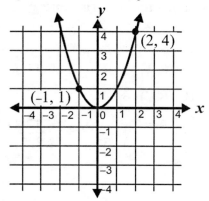

Find the slope: $\dfrac{4-1}{2-(-1)} = \dfrac{3}{3} = 1$

This value can change depending on the points you choose, so the slope of a quadratic equation is variable.

Example 18:   Absolute Value Function $y = |x|$

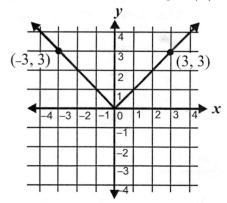

Find the slope: $\dfrac{3-3}{3-(-3)} = \dfrac{0}{6} = 0$

This, like the quadratic function, is constantly changing depending on the points you choose. It has a variable rate of change.

Graph and tell whether the slope is constant (c) or variable (v).

1. $f(x) = 3x + 2$

2. $f(x) = -x^2 + 4$

3. $f(x) = |x| + 7$

4. $f(x) = \frac{1}{2}x^2 - 1$

5. $f(x) = \sqrt{x}$

2.7   Parent Functions

The simplest function is a straight line, one in which the equation that represents it never has variables raised to a power other than 1. The simplest linear function, $y = x$, is called the **linear parent function**. Another way of saying this is that if the graph of any function is a straight line, then its parent function is $y = x$.

Example 19:   What is the parent function of the following graph?

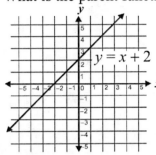

Since the graph is a straight line, the parent function is $y = x$, the linear parent function.

If an equation can be written in the form $y = ax^2 + bx + c$, where a, b, and c are constants and $a \neq 0$, then it is a **quadratic equation**. If a quadratic can be written in this form, then its **quadratic parent function** is $y = x^2$. Another way of stating this is that if the graph of any function is a parabola as shown, then its parent function is $y = x^2$.

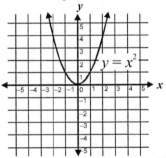

Example 20:   What is the parent function of this graph?

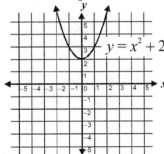

Since the graph is a parabola, its parent function is the quadratic parent function $y = x^2$. We also will notice that the equation is $y = x^2 + 2$. By the definition, this equation is quadratic, making its parent function $y = x^2$.

Name the parent function of the following equations.

1. $y = -\frac{1}{2}x + 5$

2. $y = 10x$

3. $y = x^2 - 4$

4. $y = -\frac{1}{3}x^2 + 8$

5. $y = 4x - 2$

6. $y = (x - 2)^2 - 1$

7. $y = x^2 + x + \frac{3}{2}$

8. $y = 17 - 4x$

2.8   Transformations of Graphs

Graph transformation occurs when a graph is altered from its original or "parent" state. Graphs can be altered in 5 ways: vertical shifts, horizontal shifts, they can be stretched, shrunk, or reflected.

$y = f(x + a) + b$

a = horizontal shift (left/right), b = vertical shift (up/down)

Shifting Upward and Downward

Example 21:   Given the graph of $f(x)$, what is the graph of function $y = f(x) + 3$?

$f(x) = x, y = x + 3$ $f(x) = x^2, y = x^2 + 3$

 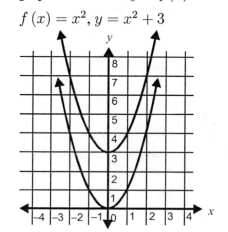

The graph of function $y = f(x) + 3$ shifts the graph of $f(x)$ up 3 units.

This transformation is called a **vertical shift**.

Shifting Left and Right

Example 22:   Given the graph of $f(x)$, what is the graph of function $y = f(x + 3)$?

$f(x) = x^2, y = (x + 3)^2$ $f(x) = |x|, y = |x + 3|$

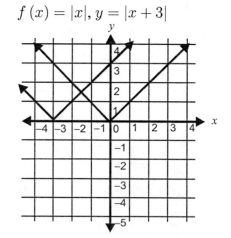

The graph of function $y = f(x + 3)$ shifts the graph of $f(x)$ left 3 units.

This transformation is called a **horizontal shift**.

Vertical Stretching and Shrinking

Example 23:   Given the graph of $f(x)$, what is the graph of function $y = 3f(x)$?

$f(x) = x^2, y = 3x^2$

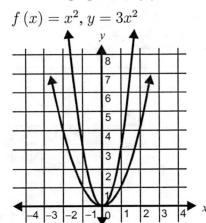

$f(x) = |x|, y = 3|x|$

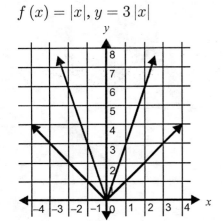

The graph of function $y = 3f(x)$ stretches the graph of $f(x)$ vertically.

This transformation is called **stretching**.

Note: If $c > 1$, then the graph of $y = cf(x)$ is the graph of $f(x)$ stretched vertically by the ratio of c. If $0 < c < 1$, then the graph of $y = cf(x)$ is the graph of $f(x)$ compressed (shrunk) vertically by the ratio of c.

Horizontal Stretching and Shrinking

Example 24:   Given the graph of $f(x)$, what is the graph of function $y = f(3x)$?

$f(x) = x^2, y = (3x)^2 = 9x^2$

$f(x) = x^2 + 1, y = 9x^2 + 1$

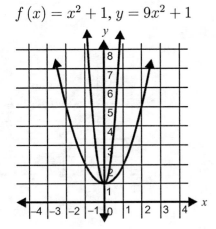

The graph of function $y = f(3x)$ shrinks the graph of $f(x)$ horizontally.

This transformation is called **shrinking**.

Note: If $c > 1$, then the graph of $y = f(cx)$ is the graph of $f(x)$ shrunk horizontally by the ratio of c. If $0 < c < 1$, then the graph of $y = f(cx)$ is the graph of $f(x)$ stretched horizontally by the ratio of c.

Reflection across the x-axis

Example 25:   Given the graph of $f(x)$, what is the graph of function $y = -f(x)$?

$f(x) = x, y = -x$ $f(x) = x^2, y = -x^2$

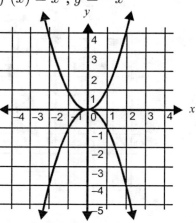

The graph of function $y = -f(x)$ reflects the graph of $f(x)$ across the x-axis.

This transformation is called a **reflection**.

Reflection across the y-axis

Example 26:   Given the graph of $f(x)$, what is the graph of function $y = f(-x)$?

$f(x) = \sqrt{x}, y = \sqrt{-x}$ $f(x) = x^3, y = (-x)^3 = -x^3$

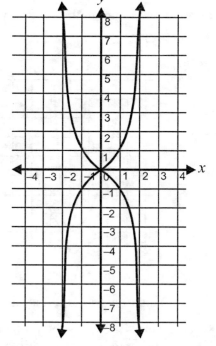

The graph of function $y = f(-x)$ reflects the graph of $f(x)$ across the y-axis.

This transformation is called a **reflection**.

Complete the following.

1. How is $y = x - 7$ different from its parent graph?

2. How does $y = 3x^2$ differ from its parent graph?

3. What is the parent graph of $y = x^2 - 12$?

4. Does $y = -f(x)$ reflect across the x or y-axis?

5. Does $y = f(-x)$ reflect across the x or y-axis?

6. How would adding 3 to $y = x^2$ change the graph?

7. How does subtracting 3 from $y = x^2$ change the graph?

8. Would $y = (x + 3)^2$ be shifted vertically or horizontally from its parent function?

9. Would $y = f(3x)$ shrink the graph of $f(x)$ horizontally or vertically?

10. What is the parent function of $y = x + 12$?

11. What is the parent function of $y = x^2 + 3x - 4$?

12. Would $y = |x| - 7$ be shifted vertically or horizontally from its parent function?

2.9   Determining Domain and Range From Graphs

The domain is all of the x values that lie on the function in the graph from the lowest x value to the highest x value. The range is all of the y values that lie on the function in the graph from the lowest y to the highest y.

Example 27:   Find the domain and range of the graph.

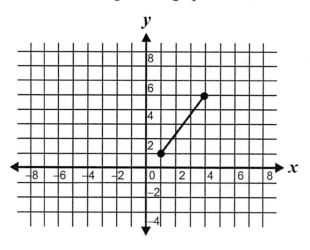

Step 1:   First find the lowest x value depicted on the graph. In this case it is 1. Then find the highest x value depicted on the graph. The highest value of x on the graph is 4. The domain must contain all of the values between the lowest x value and the highest x value. The easiest way to write this is $1 \leq \text{Domain} \leq 4$ or $1 \leq x \leq 4$.

Step 2:   Perform the same process for the range, but this time look at the lowest and highest y values. The answer is $1 \leq \text{Range} \leq 5$ or $1 \leq y \leq 5$.

Find the domain and range of each graph below. Write your answers on the line provided.

1.

2.

_____   _____

3.

_____

4.

_____

5.

_____

6.

_____

7.

_____

8.

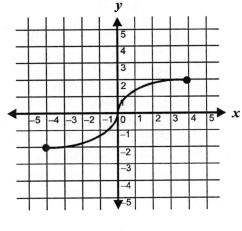

_____

2.10   Reading Non-Linear Graphs

In addition to continuity, there are a number of features that you can identify from the graph of a non-linear function.

Zeros are the points where $y = 0$. The zeros are the function's x-intercepts.

The **maximum** (plural: maxima) is the highest y value that a function reaches.

The **minimum** (plural: minima) is the lowest y value that a function reaches.

The **range** is all of the y values that are defined by the function.

Similarly, the **domain** is all of the x values defined by the function.

Finally, you can determine the **rate of increase or decrease** between two points by finding the rise over the run (slope).

Example 28:   What are the zeros, maximum, minimum, range, and domain of the function below? At what point is the function discontinuous? Also, what is the rate of increase between $x = 0$ and $x = 2$.

$$y = \begin{cases} 2x & \text{where} \quad 0 \le x \le 2 \\ \dfrac{x}{2} & \text{where} \quad 2 < x \le 4 \end{cases}$$

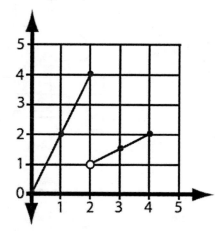

If the graph contains a point, it is shown with a closed circle.

If the graph does not contain a point, like point $(2, 1)$, you must use a open circle.

Zeros: Where does $y = 0$? Where does the graph intersect the x-intercept? Only at $x = 0$.

Maximum: Highest point is $(2, 4)$. Since the maximum point of the graph is always the largest y value that the graph reaches, 4 is the maximum.

Minimum: Lowest point is $(0, 0)$. Since the minimum point of the graph is always the smallest y value that the graph reaches, 0 is the minimum.

Range: Maximum $-$ Minimum $= 4 - 0 = 4$

Domain: Highest x $-$ Lowest $x = 4 - 0 = 4$

The function is discontinuous at $x = 2$ because this is the point where the graph is not connected.

Rate of increase between $x = 0$ and $x = 2$: At $x = 0$, $y = 0$. At $x = 2$, $y = 4$. Therefore, the rise is $4 - 0 = 4$, and the run is $2 - 0 = 2$.

Slope = rise over run $= \frac{4}{2} = 2$.

Use the following graph of the absolute value function $y = |x - 2|$ **for questions 1 and 2.**

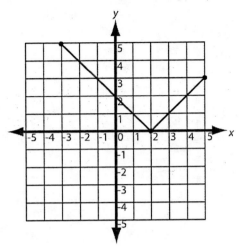

1. According to the graph, what is the minimum value of the function?

(A) -3

(B) 0

(C) 2

(D) 5

2. At what point is this graph discontinuous?

(A) $(-3, 5)$

(B) $(0, 2)$

(C) $(2, 0)$

(D) None of the above. The graph is continuous.

The following graph show a piecewise function.

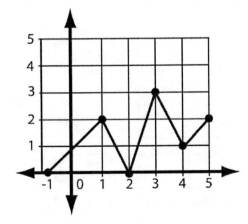

3. How many zeros does this function have?

(A) -1

(B) 0

(C) 1

(D) 2

4. What are the range and domain of the function?

(A) range: $0 \leq y \leq 3$, domain: $-3 \leq x \leq 5$

(B) range: $-3 \leq x \leq 3$, domain: $0 \leq y \leq 5$

(C) range: $-1 \leq x \leq 5$, domain: $0 \leq y \leq 3$

(D) range: $0 \leq y \leq 3$, domain: $-1 \leq x \leq 5$

Use the following graph of a step function for questions 5 and 6.

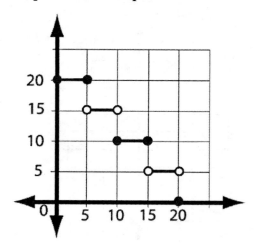

5. At how many points is this graph discontinuous?

   (A) 4
   (B) 8
   (C) 6
   (D) 2

6. What is the rate of increase of the function between 10 and 15?

   (A) -5
   (B) 0
   (C) 1
   (D) 5

Use the following graph of a non-linear function for questions 7 and 8.

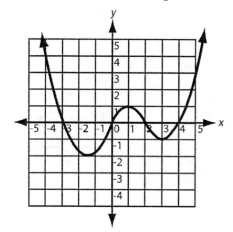

7. What are the zeros of this function?

   (A) 0
   (B) $-3, 0, 2, 4$
   (C) $-3, -1, 1, 3$
   (D) $0, 2, 4$

8. Which of the following is true of this function?

   (A) The minimum value is -2.
   (B) The graph is discontinuous at $x = 0$.
   (C) The maximum value is 5.
   (D) The graph has only 2 zeros.

2.11   Finding the Intercepts of a Line

The x-intercept is the point where the graph of a line crosses the x-axis. The y-intercept is the point where the graph of a line crosses the y-axis.

To find the x-intercept, set $y = 0$

To find the y-intercept, set $x = 0$

Example 29:   Find the x- and y-intercepts of the line $6x + 2y = 18$

Step 1:   To find the x-intercept, set $y = 0$.

$$
\begin{aligned}
6x + 2(0) &= 18 \\
\frac{6x}{6} &= \frac{18}{6} \\
x &= 3
\end{aligned}
$$

The x-intercept is at the point $(3, 0)$.

Step 2:   To find the y-intercept, set $x = 0$.

$$
\begin{aligned}
6(0) + 2y &= 18 \\
\frac{2y}{2} &= \frac{18}{2} \\
y &= 9
\end{aligned}
$$

The y-intercept is at the point $(0, 9)$.

Step 3:   You can now use the two intercepts to graph the line.

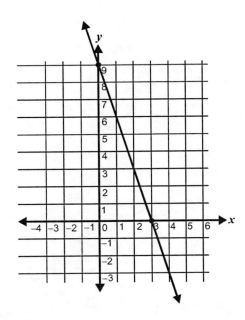

For each of the following equations, find both the x and the y intercepts of the line. For extra practice, draw each of the lines on graph paper.

1. $8x - 2y = 8$

2. $4x + 8y = 16$

3. $3x + 3y = 9$

4. $x - 2y = -5$

5. $8x + 4y = 32$

6. $3x - 4y = 12$

7. $-3x - 3y = 6$

8. $-6x + 2y = 18$

9. $4x - 2y = -4$

10. $-5x - 3y = 15$

11. $3x - 6y = -12$

12. $6x + 3y = 9$

13. $-2x - 6y = 18$

14. $2x + 3y = -6$

15. $-3x + 8y = 12$

16. $5x + 10y = 30$

17. $x - 3y = -2$

18. $-3x - 4y = 6$

19. $7x + 2y = 14$

20. $5x + 6y = 30$

21. $7x - 2y = 14$

22. $5x + 5y = 25$

23. $-14x + 4y = 28$

24. $22x - 11y = -44$

25. $-3x - 6y = -9$

26. $4x + 6y = 24$

27. $10x - 5y = 40$

28. $30x + 15y = 60$

29. $-9x + 18y = 54$

30. $8x + 7y = 56$

Chapter 2 Review

1. Graph the solution set for the linear equation: $x - 3 = y$.

2. What is the slope of the line that passes through the points $(5, 3)$ and $(6, 1)$?

3. What is the slope of the line that passes through the points $(-1, 4)$ and $(-6, -2)$?

4. What is the x-intercept for the following equation? $6x - y = 30$

5. What is the y-intercept for the following equation? $4x + 2y = 28$

6. What is the slope of the line $y = -\frac{1}{2}x + 3$?

7. What is the x-intercept of the line $y = 5x + 6$?

8. What is the y-intercept of the line $y - \frac{2}{3}x + 3 = 0$?

9. Graph the equation $y = -\frac{1}{2}x^2 + 1$.

10. What is the name of the graph described by the equation $y = 2x^2 - 1$?

Graph the equations below on a Cartesian plane.

11. $y = -x^2 + 4$

12. $y = 3x^2 - 1$

13. $y = 4 - x^2$

14. $y = 3x^2 + 6$

Graph each equation and state whether the slope is constant(c) or varying(v).

15. $f(x) = x - 6$

16. $f(x) = \frac{1}{2}x + 6$

17. $f(x) = \frac{1}{2}x^2 + 6$

18. $f(x) = \frac{1}{2}|x| + 6$

Use what you know about graph transformations to answer the following questions.

19. What is the parent graph of $y = 3x + 2$?

20. True or False: $y = f(-x)$ is reflected across the y-axis.

21. True or False: $y = 3f(x)$ shrinks a graph horizontally.

22. Write the equation of $y = x^2$ shifted down two units.

23. Write the equation of $y = x$ shifted left six units.

24. True or false: $y = f(2x)$ shrinks a graph horizontally.

Graph each function.

25. $f(x) = -|2x|$

26. $f(x) = \frac{1}{2}x^2$

27. $f(x) = |-6x|$

28. $f(x) = \frac{1}{7}x$

29. $f(x) = |4x|$

30. $f(x) = -3\sqrt{x}$

31. $f(x) = \frac{1}{4}x^3$

32. $f(x) = -\frac{1}{8}\sqrt{x}$

Answer the following questions about the nonlinear graphs.

Notre Pere College Preparatory Academy uses a 4.0 scale for its grades, to emphasize its college preparatory curriculum. Mr. Beau teaches the senior economics class, and he uses the grading scale graphed below.

33. MaryLee scored an average of 83% in the course. What was her official 4.0 scale grade?

34. What the domain and range of this function?

Use the graph below for questions 35–36.

35. What were the high and low temperatures (maximum and minimum) on Christmas Day in Hagerstown?

36. What was the approximate rate of increase between 12 PM and 4 PM?

Chapter 2 Test

1. The following graph depicts the height of a projectile as a function of time.

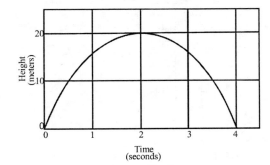

What is the domain (D) of this function?

(A) 0 meters $\leq$ D $\leq$ 20 meters
(B) 4 seconds $\leq$ D $\leq$ 20 meters
(C) 20 meters $\leq$ D $\leq$ 4 seconds
(D) 0 seconds $\leq$ D $\leq$ 4 seconds

2. What is the x-intercept of the following linear equation $3x + 4y = 12$?

(A) $(0, 3)$
(B) $(3, 0)$
(C) $(0, 4)$
(D) $(4, 0)$

3. Which of the following equations is represented by the graph?

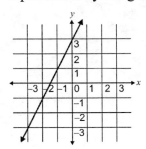

(A) $y = -3x + 3$

(B) $y = -\frac{1}{3}x + 3$

(C) $y = 3x - 3$

(D) $y = 2x + 3$

4. Which of the following is the graph of the equation $y = x - 3$?

(A)

(B)

(C)

(D)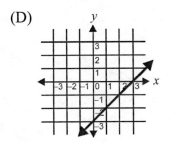

5. What function would have a variable rate of change?

(A) $f(x) = 2x^2 + 1$
(B) $f(x) = 3x - 1$
(C) $f(x) = 6x + 4$
(D) $f(x) = 9x - \frac{1}{2}$

6. What is the x-intercept and y-intercept for the equation $x + 2y = 6$?

(A) x-intercept $= (0, 6)$
y-intercept $= (3, 0)$

(B) x-intercept $= (4, 1)$
y-intercept $= (2, 2)$

(C) x-intercept $= (0, 6)$
y-intercept $= (0, 3)$

(D) x-intercept $= (6, 0)$
y-intercept $= (0, 3)$

7. Look at the graphs below. Which of the following statements is false?

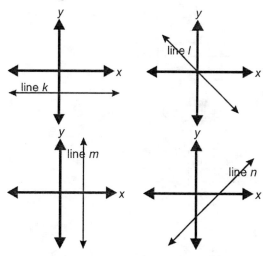

(A) The slope of line k is undefined.
(B) The slope of line l is negative.
(C) The slope of line m is undefined.
(D) The slope of line n is positive.

8. What is the rate of change of $f(x) = 3x - 6$?

(A) -6

(B) 3

(C) $\frac{1}{3}$

(D) Not enough information

9. Which of the following graphs represents $y = 2x^2$?

(A)

(B)

(C)

(D)

10. What function would have a constant rate of change?

(A) $f(x) = 2x^2 + 1$
(B) $f(x) = 7x^3 + 1$
(C) $f(x) = 403x + 6$
(D) $f(x) = 17x^4 + 4$

11. What type of rate change does $f(x) = 2x^2 + 4$ have?

(A) variable
(B) constant
(C) increasing
(D) decreasing

12. In which direction is $y = x + 6$ shifted from its parent function?

(A) right
(B) up
(C) down
(D) none of the above

13. If you have the equation $y = f(x)$, and it is changed to $y = f(-x)$, then the graph is reflected. The graph is reflected over the

(A) y-axis.
(B) x-axis.
(C) origin.
(D) line $y = x$.

14. In the equation $y = 6f(x)$, 6 would _____ the graph.

(A) vertically stretch
(B) vertically shrink
(C) vertically shift
(D) reflect

15. In the equation $y = f(6x)$, 6 would _____ the graph.

(A) horizontally stretch
(B) horizontally shrink
(C) horizontally shift
(D) reflect

16. In the equation $y = f(x+3)$ would _____ the graph _____ units _____.

(A) stretch, 3, right
(B) stretch, 3, left
(C) shift, 3, right
(D) shift, 3, left

17. $f(x) = \dfrac{1}{x}$ does not exist at what point?

(A) $x = 1$
(B) $x = -1$
(C) $x = 0$
(D) x exists everywhere

18. Which is the graph of $f(x) = \dfrac{1}{x}$?

(A)

(B)

(C)

(D)

Chapter 3
Roots

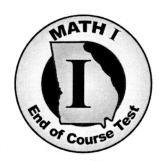

This chapter covers the following Georgia Performance Standards:

MM1A	Algebra	MM1A2a MM1A2b

3.1   Square Root

Just as working with exponents is related to multiplication, finding square roots is related to division. In fact, the sign for finding the square root of a number looks similar to a division sign. The best way to learn about square roots is to look at examples.

Examples: This is a square root problem: $\sqrt{64}$
It is asking, "What is the square root of 64?"
It means, "What number multiplied by itself equals 64?"
The answer is 8. $8 \times 8 = 64$.

Find the square roots of the following numbers.

1. $\sqrt{49}$ 4. $\sqrt{16}$ 7. $\sqrt{100}$ 10. $\sqrt{36}$ 13. $\sqrt{64}$

2. $\sqrt{81}$ 5. $\sqrt{121}$ 8. $\sqrt{289}$ 11. $\sqrt{4}$ 14. $\sqrt{9}$

3. $\sqrt{25}$ 6. $\sqrt{625}$ 9. $\sqrt{196}$ 12. $\sqrt{900}$ 15. $\sqrt{144}$

3.2   Simplifying Square Roots

Square roots can sometimes be simplified even if the number under the square root is not a perfect square. One of the rules of roots is that if a and b are two positive real numbers, then it is always true that $\sqrt{a \cdot b} = \sqrt{a} \cdot \sqrt{b}$. You can use this rule to simplify square roots.

Example 1: $\sqrt{100} = \sqrt{4 \cdot 25} = \sqrt{4} \cdot \sqrt{25} = 2 \cdot 5 = 10$

Example 2: $\sqrt{200} = \sqrt{100 \cdot 2} = 10\sqrt{2}$ ⟵ Means 10 multiplied by the square root of 2

Example 3: $\sqrt{160} = \sqrt{10 \cdot 16} = 4\sqrt{10}$

Simplify.

1. $\sqrt{98}$ 3. $\sqrt{50}$ 5. $\sqrt{8}$ 7. $\sqrt{48}$ 9. $\sqrt{54}$ 11. $\sqrt{72}$ 13. $\sqrt{90}$ 15. $\sqrt{18}$

2. $\sqrt{600}$ 4. $\sqrt{27}$ 6. $\sqrt{63}$ 8. $\sqrt{75}$ 10. $\sqrt{40}$ 12. $\sqrt{80}$ 14. $\sqrt{175}$ 16. $\sqrt{20}$

3.3   Extracting Monomial Roots

When finding the roots of monomial expressions, you must first divide the monomial expression into separate parts. Then, simplify each part of the expression.

Note: To find the square root of any variable raised to a positive exponent, simply divide the exponent by 2. For example, $\sqrt{y^{10}} = y^5$.

Example 4: $\sqrt{25x^6y^4z^6}$

Step 1:   Break each component apart. $\left(\sqrt{25}\right)\left(\sqrt{x^6}\right)\left(\sqrt{y^4}\right)\left(\sqrt{z^6}\right)$

Step 2:   Solve for each component. $\left(\sqrt{25}=5\right)\left(\sqrt{x^6}=x^3\right)\left(\sqrt{y^4}=y^2\right)\left(\sqrt{z^6}=z^3\right)$

Step 3:   Recombine the simplified expressions. $(5)\left(x^3\right)\left(y^2\right)\left(z^3\right) = 5x^3y^2z^3$

Simplify the problems below.

1. $\sqrt{4a^2b^4c^8}$

2. $\sqrt{49h^{24}i^6j^4}$

3. $\sqrt{121p^{20}q^{24}r^6}$

4. $\sqrt{36a^{28}b^{10}c^6}$

5. $\sqrt{144t^{44}u^{30}v^2}$

6. $\sqrt{36k^6l^{26}m^{20}}$

7. $\sqrt{25s^6t^{24}u^{32}}$

8. $\sqrt{81x^6y^{20}z^{44}}$

9. $\sqrt{49a^6b^4c^8}$

10. $\sqrt{169u^{12}v^{24}w^{28}}$

11. $\sqrt{64x^{44}y^{20}z^6}$

12. $\sqrt{4d^{30}e^{54}f^{10}}$

13. $\sqrt{f^4g^6h^{26}}$

14. $\sqrt{900l^{50}m^{26}n^4}$

15. $\sqrt{400g^{40}h^{26}i^{36}}$

16. $\sqrt{25a^{54}b^6c^{46}}$

17. $\sqrt{16j^{24}k^8l^{20}}$

18. $\sqrt{9q^4r^{20}s^{34}}$

3.4   Monomial Roots with Remainders

Monomial roots which are not easily simplified under the square root symbol will also sometimes be encountered. Powers may be raised to odd numbers. In addition, the coefficients may not be perfect squares. Follow the example below to understand how to simplify these types of problems.

Example 5:   Simplify $\sqrt{40x^7y^{11}z^{23}}$

Step 1:   Begin by simplifying the coefficient. $\sqrt{40} = \left(\sqrt{4}\right)\left(\sqrt{10}\right)$, $\sqrt{4} = 2$, so $\sqrt{40} = 2\sqrt{10}$

Step 2:   Simplify the variable with exponents.

$$\sqrt{x^7} = \left(\sqrt{x^6}\right)\left(\sqrt{x}\right), \sqrt{x^6} = x^3, \text{ so } \sqrt{x^7} = x^3\sqrt{x}$$

$$\sqrt{y^{11}} = \left(\sqrt{y^{10}}\right)\left(\sqrt{y}\right), \sqrt{y^{10}} = y^5, \text{ so } \sqrt{y^{11}} = y^5\sqrt{y}$$

$$\sqrt{z^{23}} = \left(\sqrt{z^{22}}\right)\left(\sqrt{z}\right), \sqrt{z^{22}} = z^{11}, \text{ so } \sqrt{z^{23}} = z^{11}\sqrt{z}$$

Step 3:   Recombine the simplified expressions. $2x^3y^5z^{11}\sqrt{10xyz}$

Simplify the following square root expressions.

1. $\sqrt{57d^{25}e^{27}f^{22}}$

2. $\sqrt{140h^{26}i^{20}j^9}$

3. $\sqrt{27x^{44}y^{42}z^9}$

4. $\sqrt{75p^{22}q^8r^{21}}$

5. $\sqrt{48k^{47}l^{27}m^3}$

6. $\sqrt{75s^{23}t^7u^{28}}$

7. $\sqrt{63a^8b^{27}c^{42}}$

8. $\sqrt{20p^3q^{44}r^{29}}$

9. $\sqrt{80a^{220}b^{20}c^{27}}$

10. $\sqrt{64m^8n^3p^{22}}$

11. $\sqrt{88r^{27}s^{22}t^{23}}$

12. $\sqrt{40g^{42}h^{25}j^{28}}$

13. $\sqrt{90v^3w^{20}x^{24}}$

14. $\sqrt{50d^7e^9f^{23}}$

15. $\sqrt{45x^{28}y^6z^{23}}$

16. $\sqrt{32a^6b^{23}c^7}$

17. $\sqrt{74j^{24}k^{27}m^7}$

18. $\sqrt{20q^{24}r^{27}s^7}$

3.5   Adding and Subtracting Roots

You can add and subtract terms with square roots only if the number under the square root sign, the radical, is the same.

Example 6: $2\sqrt{2} + 3\sqrt{2} = 5\sqrt{2}$

Example 7: $12\sqrt{7} - 3\sqrt{7} - 9\sqrt{7}$

Or, look at the following examples where you can simplify the square roots and then add or subtract.

Example 8: $2\sqrt{25} + \sqrt{36}$

   Step 1:   Simplify. You know that $\sqrt{25} = 5$, and $\sqrt{36} = 6$ so the problem simplifies to $2(5) + 6$

   Step 2:   Solve: $2(5) + 6 = 10 + 6 = 16$

Example 9: $2\sqrt{72} - 3\sqrt{2}$

   Step 1:   Simplify what you know. $\sqrt{72} = \sqrt{36 \cdot 2} = 6\sqrt{2}$

   Step 2:   Substitute $6\sqrt{2}$ for $\sqrt{72}$ simplify.
   $2(6)\sqrt{2} - 3\sqrt{2} = 12\sqrt{2} - 3\sqrt{2} = 9\sqrt{2}$

Simplify the following addition and subtraction problems.

1. $3\sqrt{5} + 9\sqrt{5}$

2. $3\sqrt{25} + 4\sqrt{16}$

3. $4\sqrt{8} + 2\sqrt{2}$

4. $3\sqrt{32} - 2\sqrt{2}$

5. $\sqrt{25} - \sqrt{49}$

6. $2\sqrt{5} + 4\sqrt{20}$

7. $5\sqrt{8} - 3\sqrt{72}$

8. $\sqrt{27} + 3\sqrt{27}$

9. $3\sqrt{20} - 4\sqrt{45}$

10. $4\sqrt{45} - \sqrt{75}$

11. $2\sqrt{28} + 2\sqrt{7}$

12. $\sqrt{64} + \sqrt{81}$

13. $5\sqrt{54} - 2\sqrt{24}$

14. $\sqrt{32} + 2\sqrt{50}$

15. $2\sqrt{7} + 4\sqrt{63}$

16. $8\sqrt{2} + \sqrt{8}$

17. $2\sqrt{8} - 4\sqrt{32}$

18. $\sqrt{36} + \sqrt{100}$

19. $\sqrt{9} + \sqrt{25}$

20. $\sqrt{64} - \sqrt{36}$

21. $\sqrt{75} + \sqrt{108}$

22. $\sqrt{81} + \sqrt{100}$

23. $\sqrt{192} - \sqrt{75}$

24. $3\sqrt{5} + \sqrt{245}$

3.6   Multiplying Roots

You can also multiply square roots. To multiply square roots, you just multiply the numbers under the radical and then simplify. Look at the examples below.

Example 10: $\sqrt{2} \times \sqrt{6}$

Step 1: $\sqrt{2} \times \sqrt{6} = \sqrt{2 \times 6} = \sqrt{12}$ Multiply the numbers under the radical.

Step 2: $\sqrt{12} = \sqrt{4 \times 3} = 2\sqrt{3}$ Simplify.

Example 11: $3\sqrt{3} \times 5\sqrt{6}$

Step 1: $(3 \times 5)\sqrt{3 \times 6} = 15\sqrt{18}$ Multiply the numbers in front of the square root, and multiply the numbers under the radical.

Step 2: $15\sqrt{18} = 15\sqrt{2 \times 9}$ Simplify.
$15 \times 3\sqrt{2} = 45\sqrt{2}$

Example 12: $\sqrt{14} \times \sqrt{42}$ For this more complicated multiplication problem, use the rule of roots that you learned on page 60, $\sqrt{a \cdot b} = \sqrt{a} \cdot \sqrt{b}$.

Step 1: $\sqrt{14} = \sqrt{7} \times \sqrt{2}$ and   Instead of multiplying 14 by 42, divide these
$\sqrt{42} = \sqrt{2} \times \sqrt{3} \times \sqrt{7}$ numbers into their roots.

$\sqrt{14} \times \sqrt{42} = \sqrt{7} \times \sqrt{2} \times \sqrt{2} \times \sqrt{3} \times \sqrt{7}$

Step 2:   Since you know that $\sqrt{7} \times \sqrt{7} = 7$ and $\sqrt{2} \times \sqrt{2} = 2$, the problem simplifies to
$(7 \times 2)\sqrt{3} = 14\sqrt{3}$

Simplify the following multiplication problems.

1. $\sqrt{5} \times \sqrt{7}$

2. $\sqrt{32} \times \sqrt{2}$

3. $\sqrt{10} \times \sqrt{14}$

4. $2\sqrt{3} \times 3\sqrt{6}$

5. $4\sqrt{2} \times 2\sqrt{10}$

6. $\sqrt{5} \times 3\sqrt{15}$

7. $\sqrt{45} \times \sqrt{27}$

8. $5\sqrt{21} \times \sqrt{7}$

9. $\sqrt{42} \times \sqrt{21}$

10. $4\sqrt{3} \times 2\sqrt{12}$

11. $\sqrt{56} \times \sqrt{24}$

12. $\sqrt{11} \times 2\sqrt{33}$

13. $\sqrt{13} \times \sqrt{26}$

14. $2\sqrt{2} \times 5\sqrt{5}$

15. $\sqrt{6} \times \sqrt{12}$

3.7 Dividing Roots

When dividing a number or a square root by another square root, you cannot leave the square root sign in the denominator (the bottom number) of a fraction. You must simplify the problem so that the square root is not in the denominator. This is also called rationalizing the denominator. Look at the examples below.

Example 13: $\dfrac{\sqrt{2}}{\sqrt{5}}$

Step 1: $\dfrac{\sqrt{2}}{\sqrt{5}} \times \dfrac{\sqrt{5}}{\sqrt{5}}$ ⟵ The fraction $\frac{\sqrt{5}}{\sqrt{5}}$ is equal to 1, and multiplying by 1 does not change the value of a number.

Step 2: $\dfrac{\sqrt{2 \times 5}}{5} = \dfrac{\sqrt{10}}{5}$ Multiply and simplify. Since $\sqrt{5} \times \sqrt{5}$ equals 5, you no longer have a square root in the denominator.

Example 14: $\dfrac{6\sqrt{2}}{2\sqrt{10}}$ In this problem, the numbers outside of the square root will also simplify.

Step 1: $\dfrac{6}{2} = 3$ so you have $\dfrac{3\sqrt{2}}{\sqrt{10}}$

Step 2: $\dfrac{3\sqrt{2}}{\sqrt{10}} \times \dfrac{\sqrt{10}}{\sqrt{10}} = \dfrac{3\sqrt{2 \times 10}}{10} = \dfrac{3\sqrt{20}}{10}$

Step 3: $\dfrac{3\sqrt{20}}{10}$ will further simplify because $\sqrt{20} = 2\sqrt{5}$, so you then have $\dfrac{3 \times 2\sqrt{5}}{10}$ which reduces to $\dfrac{3\sqrt{5}}{5}$.

Simplify the following division problems.

1. $\dfrac{9\sqrt{3}}{\sqrt{5}}$

2. $\dfrac{16}{\sqrt{8}}$

3. $\dfrac{24\sqrt{10}}{12\sqrt{3}}$

4. $\dfrac{\sqrt{121}}{\sqrt{6}}$

5. $\dfrac{\sqrt{40}}{\sqrt{90}}$

6. $\dfrac{33\sqrt{15}}{11\sqrt{2}}$

7. $\dfrac{\sqrt{32}}{\sqrt{12}}$

8. $\dfrac{\sqrt{11}}{\sqrt{5}}$

9. $\dfrac{\sqrt{2}}{\sqrt{6}}$

10. $\dfrac{2\sqrt{7}}{\sqrt{14}}$

11. $\dfrac{5\sqrt{2}}{4\sqrt{8}}$

12. $\dfrac{4\sqrt{21}}{7\sqrt{7}}$

13. $\dfrac{9\sqrt{22}}{2\sqrt{2}}$

14. $\dfrac{\sqrt{35}}{2\sqrt{14}}$

15. $\dfrac{\sqrt{40}}{\sqrt{15}}$

16. $\dfrac{\sqrt{3}}{\sqrt{12}}$

Chapter 3 Review

Simplify the following square root expressions.

1. $\sqrt{50}$

2. $\sqrt{44}$

3. $\sqrt{12}$

4. $\sqrt{18}$

5. $\sqrt{8}$

6. $\sqrt{48}$

7. $\sqrt{75}$

8. $\sqrt{200}$

9. $\sqrt{32}$

10. $\sqrt{20}$

11. $\sqrt{63}$

12. $\sqrt{80}$

13. $\sqrt{36r^6 s^8 t^2}$

14. $\sqrt{40g^3 h^6 j^7}$

15. $\sqrt{18m^5 n^3 p^7}$

16. $\sqrt{75a^7 b^2 c^9}$

Simplify the following square root problems.

17. $5\sqrt{27} + 7\sqrt{3}$

18. $\sqrt{40} - \sqrt{10}$

19. $\sqrt{64} + \sqrt{81}$

20. $8\sqrt{50} - 3\sqrt{32}$

21. $14\sqrt{5} + 8\sqrt{80}$

22. $\sqrt{63} \times \sqrt{28}$

23. $\dfrac{\sqrt{56}}{\sqrt{35}}$

24. $\sqrt{8} \times \sqrt{50}$

25. $\dfrac{\sqrt{20}}{\sqrt{45}}$

26. $5\sqrt{40} \times 3\sqrt{20}$

27. $2\sqrt{48} - \sqrt{12}$

28. $\dfrac{2\sqrt{5}}{\sqrt{30}}$

29. $\dfrac{3\sqrt{22}}{2\sqrt{3}}$

30. $\sqrt{72} \times 3\sqrt{27}$

Chapter 3 Test

1. Simplify: $\sqrt{135}$

   (A) $3\sqrt{15}$

   (B) $\sqrt{72}$

   (C) $9\sqrt{15}$

   (D) $\sqrt{9} \times \sqrt{15}$

2. Express $\dfrac{\sqrt{20}}{\sqrt{35}}$ in simplest form.

   (A) $\dfrac{2\sqrt{7}}{7}$

   (B) $\dfrac{2}{\sqrt{7}}$

   (C) $\dfrac{2\sqrt{5}}{\sqrt{7}}$

   (D) $\dfrac{4}{7}$

3. Simplify: $\sqrt{44} \cdot 2\sqrt{33}$

   (A) $2\sqrt{77}$

   (B) $44\sqrt{3}$

   (C) $22\sqrt{7}$

   (D) $22\sqrt{12}$

4. Simplify: $\sqrt{12x^3y^4z^7}$

   (A) $2xyz^3\sqrt{3xyz}$

   (B) $xy^2z^3\sqrt{12xz}$

   (C) $2xy^2z^3\sqrt{3xz}$

   (D) $2xy^2z^3\sqrt{3}$

5. Simplify: $\sqrt{45} \times \sqrt{27}$

   (A) $3\sqrt{15}$

   (B) $\sqrt{72}$

   (C) $9\sqrt{15}$

   (D) $\sqrt{9} \times \sqrt{15}$

6. Simplify: $\dfrac{3\sqrt{12}}{2\sqrt{3}}$

   (A) 3

   (B) $\dfrac{3\sqrt{4}}{2}$

   (C) $\dfrac{6}{\sqrt{6}}$

   (D) $3\sqrt{3}$

7. Simplify the following by rationalizing the denominator.

   $$\dfrac{\sqrt{3}}{\sqrt{15}}$$

   (A) $\dfrac{\sqrt{5}}{5}$

   (B) $\dfrac{1}{\sqrt{5}}$

   (C) $\dfrac{\sqrt{45}}{15}$

   (D) $\dfrac{3\sqrt{5}}{15}$

8. Simplify: $\sqrt{64f^6g^7h^5}$

   (A) $8f^3g^3h^2\sqrt{gh}$
   (B) $8f^3g^3h^2$
   (C) $f^3g^3h^2\sqrt{64gh}$
   (D) $8f^3g^3h^2\sqrt{g}$

Chapter 4
Polynomials

This chapter covers the following Georgia Performance Standards:

MM1A	Algebra	MM1A2c
		MM1A2d
		MM1A2g

Polynomials are algebraic expressions which include **monomials** containing one term, **binomials** which contain two terms, and **trinomials**, which contain three terms. Expressions with more than three terms are called **polynomials**. **Terms** are separated by plus and minus signs.

EXAMPLES

Monomials	Binomials	Trinomials	Polynomials
$5f$	$5t + 20$	$x^2 + 4x + 3$	$x^3 - 3x^4 + 3x - 20$
$3x^3$	$20 - 8g$	$7x^4 - 6x - 2$	$p^5 + 4p^3 + p^4 + 20p - 7$
$5g^4$	$7x^4 + 8x$	$y^5 + 27y^4 + 200$	
4	$6x^3 - 9x$		

4.1 Adding and Subtracting Monomials

Two **monomials** are added or subtracted as long as the **variable and its exponent** are the **same**. This is called combining like terms. Use the same rules you used for adding and subtracting integers

Example 1: $5x + 7x = 12x$ $\begin{array}{r} 3x^5 \\ -9x^5 \\ \hline -6x^5 \end{array}$ $4x^4 - 20x^4 = -16x^4$ $\begin{array}{r} 7y \\ +4y \\ \hline 11y \end{array}$ $6y^3 - 7y^3 = -y^3$

Remember: When the integer in front of the variable is "1", the one is usually not written. $1x^4$ is the same as x^4, and $-1x$ is the same as $-x$.

Add or subtract the following monomials.

1. $4x^4 + 7x^4 =$
2. $7t + 9t =$
3. $20y^3 - 4y^3 =$
4. $6g - 9g =$
5. $8y^4 + 9y^4 =$

6. $s^7 + s^7 =$
7. $-4x - 5x =$
8. $5w^4 - w^4 =$
9. $z^5 + 20z^5 =$
10. $-k + 4k =$

11. $3x^4 - 7x^4 =$
12. $20t + 4t =$
13. $-8v^3 + 20v^3 =$
14. $-4x^3 + x^3 =$
15. $20y^5 - 7y^5 =$

16. $\begin{array}{r} y^5 \\ +4y^5 \\ \hline \end{array}$ 18. $\begin{array}{r} 9t^4 \\ +8t^4 \\ \hline \end{array}$ 20. $\begin{array}{r} 7w^4 \\ +9w^4 \\ \hline \end{array}$ 22. $\begin{array}{r} -7z \\ +20z \\ \hline \end{array}$ 24. $\begin{array}{r} 8t^3 \\ -6t^3 \\ \hline \end{array}$

17. $\begin{array}{r} 5x^3 \\ -20x^3 \\ \hline \end{array}$ 19. $\begin{array}{r} -4y \\ -5y \\ \hline \end{array}$ 21. $\begin{array}{r} 22t^3 \\ -5t^3 \\ \hline \end{array}$ 23. $\begin{array}{r} 5w^7 \\ +w^7 \\ \hline \end{array}$ 25. $\begin{array}{r} 3x \\ +9x \\ \hline \end{array}$

4.2   Adding Polynomials

When adding **polynomials,** make sure the exponents and variables are the same on the terms you are combining. The easiest way is to put the terms in columns with **like exponents** under each other. Each column is added as a separate problem. Fill in the blank spots with zeros if it helps you keep the columns straight. You never carry to the next column when adding polynomials.

Example 2:   Add $3x^4 + 25$ and $7x^4 + 4x$

$$\begin{array}{r} 3x^4 + 0x + 25 \\ (+)\, 7x^4 + 4x + 0 \\ \hline 10x^4 + 4x + 25 \end{array}$$

Example 3: $(5x^3 - 4x) + (-x^3 - 5)$

$$\begin{array}{r} 5x^3 - 4x + 0 \\ (+) - x^3 + 0x - 5 \\ \hline 4x^3 - 4x - 5 \end{array}$$

Add the following polynomials.

1. $y^4 + 3y + 4$ and $4y^4 + 5$
2. $(7y^4 + 5y - 6) + (4y^4 - 7y + 9)$
3. $-4x^4 + 7x^3 + 5x - 2$ and $3x^4 - x + 4$
4. $-p + 5$ and $7p^4 - 4p + 4$
5. $(w - 4) + (w^4 + 4)$
6. $5t^4 - 7t - 8$ and $9t + 4$
7. $t^5 + t + 9$ and $4t^3 + 5t - 5$
8. $(s^4 + 3s^3 - 4) + (-4s^3 + 5)$
9. $(-v^4 + 8v - 9) + (5v^3 - 6v + 5)$
10. $6m^4 - 4m + 20$ and $m^4 - m - 9$
11. $-x + 5$ and $3x^4 + x - 4$
12. $(9t^4 + 3t) + (-8t^4 - t + 5)$
13. $(3p^5 + 4p^4 - 2) + (-7p^4 - p + 9)$

14. $20s^4 + 24s^3 + 4s$ and $s^4 + s^3 + s$
15. $(-20b^4 + 8b + 4) + (-b^4 + 6b + 20)$
16. $27c^4 - 22c + 7$ and $-8c^4 + 3c - 20$
17. $4c^4 + 7c^3 + 3$ and $5c^4 + 4c^3 + 2$
18. $3x^4 + -25x^3 + 27$ and $8x^3 - 24$
19. $(-x^4 + 4x - 5) + (3x^4 - 3)$
20. $(y^4 - 22y + 20) + (-23y^4 + 7y - 5)$
21. $3d^7 - 5d^3 + 8$ and $4d^5 - 4d^3 - 4$
22. $(6t^7 - t^3 + 28) + (5t^7 + 8t^3)$
23. $5p^4 - 9p + 20$ and $-p^4 - 3p - 7$
24. $40b^3 + 27b$ and $-5b^4 - 7b + 25$
25. $(-4w + 22) + (w^3 + w - 5)$
26. $(47z^4 + 23z + 9) + (z^4 - 4z - 20)$

4.3   Subtracting Polynomials

When you subtract polynomials, it is important to remember to change all the signs in the sub-tracted polynomial (the subtrahend) and then add.

Example 4: $(5y^4 + 9y + 20) - (4y^4 + 6y - 5)$

Step 1:   Copy the subtraction problem into vertical form. Make sure you line up the terms with like exponents under each other just like you did for adding polynomials.

$$5y^4 + 9y + 20$$
$$\underline{(-)\, 4y^4 + 6y - 5}$$

Step 2:   Change the subtraction sign to addition and all the signs of the subtracted polynomial to the opposite sign. The bottom polynomial in the problem becomes $-4y^4 - 6y + 5$.

Step 3:   Add: $5y^4 + 9y + 20$
$$\underline{(+)\, -4y^4 - 6y + 5}$$
$$y^4 + 3y + 25$$

Subtract the following polynomials.

1. $(4x^4 + 7x + 4) - (x^4 + 3x + 2)$

2. $(9y - 5) - (5y + 3)$

3. $(-5t^4 + 22t^3 + 3) - (5t^4 - t^3 - 7)$

4. $(-3w^4 + 20w - 7) - (-7w^4 - 7)$

5. $(6a^7 - a^3 + a) - (8a^7 + a^4 - 3a)$

6. $(25c^5 + 40c^4 + 20) - (8c^5 + 7c^4 + 24)$

7. $(7x^4 - 20x) - (-8x^4 + 5x + 9)$

8. $(-9y^4 + 24y^3 - 20) - (3y^3 + y + 20)$

9. $(-3h^4 - 8h + 8) - (7h^4 + 5h + 20)$

10. $(20k^3 - 9) - (k^4 - 5k^3 + 7)$

11. $(x^4 - 7x + 20) - (6x^4 - 7x + 8)$

12. $(24p^4 + 5p) - (20p - 4)$

13. $(-4m - 9) - (6m + 4)$

14. $(4y^4 + 23y^3 - 9y) - (5y^4 + 4y^3 - 8y)$

15. $(8g + 3) - (g^4 + 5g - 7)$

16. $(-9w^3 + 5w) - (-5w^4 - 20w^3 - w)$

17. $(x^4 + 24x^3 - 20) - (4x^4 + 3x^3 + 2)$

18. $(4a^4 + 4a + 4) - (-a^4 + 3a + 3)$

19. $(c + 220) - (3c^4 - 8c + 4)$

20. $(-6v^4 + 24v) - (3v^4 + 4v + 6)$

21. $(3b^4 + 5b^3 + 7) - (8b^3 - 9)$

22. $(7x^4 + 27x^3 - 5) - (-5x^4 + 5x^3)$

23. $(9y^4 - 4) - (22y^4 - 4y - 3)$

24. $(-z^4 - 7z - 9) - (3z^4 - 7z + 7)$

సింధూర

12 Rooms in Saddlebrook Creek Drive

1. Sneha & Siddu

2) Akhila &

A subtraction of polynomials problem may be stated in sentence form. Study the examples below.

Example 5:   Subtract $-7x^3 + 5x - 3$ from $3x^3 + 5x^4 - 6x$.

Step 1:   Copy the problem in columns with terms with the same exponent and variable under each other. Notice the second polynomial in the sentence will be the top polynomial of the problem.

$$\begin{array}{l} 3x^3 + 5x^4 - 6x \\ \underline{(-) - 7x^3 \qquad\ + 5x - 3} \end{array}$$

Since this is a subtraction problem, change all the signs of the terms in the bottom polynomial. Then add.

$$\begin{array}{l} 3x^3 + 5x^4 - 6x \\ \underline{(+)\, 7x^3 \qquad\ - 5x + 3} \\ 10x^3 + 5x^4 - 11x + 3 \end{array}$$

Example 6:   From $6y^4 + 4$ subtract $5y^4 - 3y + 9$

In a problem phrased like this one, the first polynomial will be on top, and the second will be on bottom. Change the signs on the bottom polynomial and then add.

$$\begin{array}{l} 6y^4 \qquad + 4 \\ \underline{(-)\, 5y^4 - 3y + 9} \end{array} \quad\longrightarrow\quad \begin{array}{l} 6y^4 \qquad + 4 \\ \underline{(+) - 5y^4 + 3y - 9} \\ y^4 + 3y - 5 \end{array}$$

Solve the following subtraction problems.

1. Subtract $3x^4 + 4x - 7$ from $7x^4 + 4$

2. From $7y^3 - 6y + 20$ subtract $9y^3 - 20$

3. From $5m^4 - 5m + 8$ subtract $4m - 3$

4. Subtract $9z^4 + 3z + 4$ from $5z^4 - 8z + 9$

5. Subtract $t^4 + 20t^3 - 7$ from $-t^4 - 4t^3 - 7$

6. Subtract $-8b^3 - 4b + 5$ from $-b^4 + b + 6$

7. From $20y^3 + 40$ subtract $7y^3 - 7$

8. From $25t^4 - 6t - 9$ subtract $5t^4 - 3t + 4$

9. Subtract $3p^4 + p - 4$ from $-8p^4 - 7p + 4$

10. Subtract $x^3 + 9$ from $-4x^4 + 3x^3 + 20$

11. Subtract $24a^4 + 20$ from $-a^4 + a^3 - 2$

12. From $6m^4 + 3m + 2$ subtract $-6m^4 - 3m$

13. From $-3z^4 - 23z^3 - 4$ subtract $-40z^3 + 40$

14. Subtract $20c^4 + 20$ from $9c^4 - 7c + 3$

15. Subtract $b^4 + b - 7$ from $7b^4 - 5b + 7$

16. Subtract $-3x - 5$ from $3x^4 + x + 20$

17. From $27y^4 + 4$ subtract $5y^4 + 3y + 8$

18. Subtract $3g^4 - 7g + 7$ from $20g^4 - 3g - 5$

19. From $-8m^4 - 9m$ subtract $3m^4 + 8$

20. Subtract $x + 2$ from $7x + 7$

21. Subtract $c^4 + c + 4$ from $-c^4 - c - 4$

22. From $6t^4 + 9t^3 - 5t + 4$ subtract $t^3 + 3t$

4.4   Multiplying Monomials

When two monomials have the **same variable**, you can multiply them. Then, add the **exponents** together. If the variable has no exponent, it is understood that the exponent is 1.

Example 7: $5x^5 \times 3x^4 = 15x^9$ $\qquad\qquad$ $4y \times 7y^4 = 28y^5$

Multiply the following monomials.

1. $6a \times 20a^7$ 8. $4d^9 \times 20d^4$ 15. $9w^8 \times w$

2. $4x^6 \times 7x^3$ 9. $6k^3 \times 7k^4$ 16. $20s^6 \times 7s^3$

3. $5y^3 \times 3y^4$ 10. $8m^7 \times m$ 17. $5d^7 \times 5d^7$

4. $20t^4 \times 4t^4$ 11. $22z \times 4z^8$ 18. $7y^4 \times 9y^6$

5. $4p^7 \times 5p^4$ 12. $3w^5 \times 6w^7$ 19. $8t^{20} \times 3t^7$

6. $20b^4 \times 9b$ 13. $5x^5 \times 7x^3$ 20. $6p^9 \times 4p^3$

7. $3c^3 \times 3c^3$ 14. $7n^4 \times 3n^3$ 21. $x^3 \times 4x^3$

When problems include negative signs, follow the rules for multiplying integers.

22. $-8s^5 \times 7s^3$ 29. $20d \times -9d^8$ 36. $-5w \times -7w^9$

23. $-6a \times -20a^7$ 30. $-3g^6 \times -4g^3$ 37. $-7y^5 \times 6y^4$

24. $5x \times -x$ 31. $-8s^5 \times 8s^3$ 38. $20x^3 \times -8x^7$

25. $-3y^4 \times -y^3$ 32. $-d^3 \times -4d$ 39. $-a^5 \times -a$

26. $-7b^4 \times 3b^7$ 33. $22p \times -4p^7$ 40. $-8k^4 \times 3k$

27. $20c^5 \times -4c$ 34. $-7x^8 \times -3x^3$ 41. $-27t^4 \times -t^5$

28. $-5t^3 \times 9t^3$ 35. $9z^5 \times 8z^5$ 42. $3x^9 \times 20x^4$

4.5 Multiplying Monomials with Different Variables

Warning: You cannot add the exponents of variables that are different.

Example 8: $(-5wx)(6w^3x^4)$

To work this problem, first multiply the whole numbers: $-5 \times 6 = -30$. Then multiply the w's: $w \times w^3 = w^4$. Last, multiply the x's: $x \times x^4 = x^5$. The answer is $-30w^4x^5$.

Multiply the following monomials.

1. $(4x^4y^4)(-5xy^3) =$

2. $(20p^3q^5)(4p^4q) =$

3. $(-3t^5v^4)(t^4v) =$

4. $(8w^3z^4)(3wz) =$

5. $(-4st^6)(-9s^4t) =$

6. $(xy^3)(5x^4y^4) =$

7. $(7y^4z)(3y^5z^4) =$

8. $(-3a^4b^4)(-5ab^3) =$

9. $(-7c^3d^4)(4c^5d^7) =$

10. $(20x^5y^4)(3x^3y) =$

11. $(6f^3g^7)(-f^3g) =$

12. $(-5a^3v^5)(9a^5v) =$

13. $(7m^9n^7)(8m^4n^5) =$

14. $(8w^7y^3)(3wy) =$

15. $(4x^5z^4)(-20x^4z^5) =$

16. $(-5a^8c^{20})(4a^4c) =$

17. $(-bd^6)(-b^4d) =$

18. $(3x^5y^4)(20x^3y^3) =$

19. $(20p^4y)(7p^7y^3) =$

20. $(-4a^8x^4)(6ax^4) =$

21. $(9c^5d^3)(-4c^4d^4) =$

Multiplying three monomials works the same way. The first one is done for you.

22. $(3st)(5s^3t^4)(4s^4t^5) = 60s^8t^{10}$

23. $(xy)(x^4y^4)(4x^3y^4) =$

24. $(4a^4b^4)(a^3b^3)(4ab) =$

25. $(5y^4z^5)(4y^3)(4z^4) =$

26. $(7cd^3)(3c^4d^4)(d^4) =$

27. $(4w^4x^3)(3x^5)(4w^3) =$

28. $(a^5d^4)(ad)(a^4d^3) =$

29. $(8x^3t)(4t^5)(x^4t^4) =$

30. $(p^4y^4)(5py)(p^3y^3) =$

31. $(5x^3y)(7xy^3)(4y^5) =$

32. $(9xy^4)(x^4y^3)(4x^4y) =$

33. $(6p^3t)(4t^3)(p^4) =$

34. $(3bc)(b^4c)(5c^3) =$

35. $(4y^5z^7)(4y^6)(y^4z^4) =$

36. $(5p^3r^3)(5r^4)(p^4r) =$

37. $(a^5z^6)(6a^4z^4)(3z^3) =$

38. $(7c^3)(6d^4)(4c^4d) =$

39. $(20s^8t^4)(3st)(s^4t^3) =$

40. $(3a^3b^5)(4b^3)(3a^4) =$

41. $(7wz)(w^3z^3)(3w^4z^3) =$

4.6   Dividing Monomials

When simplifying monomial fractions with exponents, all exponents need to be positive. If there are negative exponents in your answer, put the base with its negative exponent below the fraction line and remove the negative sign. Two variables that are alike should not appear in both the denominator and numerator of a simplified expression. If you have the same variable in both the denominator and numerator of the fraction, the expression is not in simplest terms.

Example 9: $\dfrac{55x^4y^4}{11x^6y^6}$

  Step 1:   Reduce the whole numbers first. $\dfrac{55}{11} = 5$

  Step 2:   Simplify the x's. $\dfrac{x^4}{x^6} = x^{4-6} = x^{-2} = \dfrac{1}{x^2}$

  Step 3:   Simplify the y's. $\dfrac{y^4}{y^6} = y^{4-6} = y^{-2} = \dfrac{1}{y^2}$

  Therefore $\dfrac{55x^4y^4}{11x^6y^6} = \dfrac{5}{x^2y^2}$

Simplify the expressions below. All answers should only have positive exponents.

1. $\dfrac{7xy^3}{x\left(4x^3\right)y^5}$

2. $\dfrac{4a^4b^7}{3a^5b^4}$

3. $\dfrac{8\left(4a^4\right)b^5}{9ab^6}$

4. $\dfrac{26\left(x^4y^5\right)^3}{40xy}$

5. $\dfrac{20a^5b^4}{7a^6b^3}$

6. $\dfrac{7\left(9x^4y^3\right)}{5\left(x^4y^4\right)^4}$

7. $\dfrac{24\left(3a^4\right)b^4}{6a^4b^4}$

8. $\dfrac{\left(6x^3y^5\right)^4}{\left(4x^7y\right)^3}$

9. $\dfrac{27a^4b^3}{3a^7b^6}$

10. $\dfrac{33x^7y^3}{44x^8y^7}$

11. $\dfrac{27\left(4a^6b^8\right)}{42a^3b^6}$

12. $\dfrac{30x^5y^4}{6\left(x^4y^4\right)^4}$

13. $\dfrac{20\left(9ab^7\right)}{40a^4b^4}$

14. $\dfrac{20x^{20}y^8}{57x^7y^3}$

15. $\dfrac{\left(a^5b^8\right)^5}{a^9b^8}$

16. $\dfrac{8\left(x^3y^7\right)}{9x^4y^5}$

17. $\dfrac{20\left(a^3b^5\right)}{5a^7b^4}$

18. $\dfrac{48\left(x^4y^7\right)^4}{42x^3y^5}$

4.7   Multiplying Monomials by Polynomials

Example 10: $-7t\left(4t^4 - 8t + 20\right)$

Step 1:   Multiply $-7t \times 4t^4 = \mathbf{-28t^5}$

Step 2:   Multiply $-7t \times -8t = \mathbf{56t^2}$

Step 3:   Multiply $-7t \times 20 = \mathbf{-140t}$

Step 4:   Arrange the answers horizontally in order: $\mathbf{-28t^5 + 56t^2 - 140t}$

Remove parentheses in the following problems.

1. $3x\left(3x^4 + 5x - 2\right)$

2. $5y\left(y^3 - 8\right)$

3. $8a^4\left(4a^4 + 3a + 4\right)$

4. $-7d^3\left(d^4 - 7d\right)$

5. $4w\left(-5w^4 + 3w - 9\right)$

6. $9p\left(p^3 - 6p + 7\right)$

7. $-20b^4\left(-4b + 7\right)$

8. $4t\left(t^4 - 5t - 20\right)$

9. $20c\left(5c^4 + 3c - 8\right)$

10. $6z\left(4z^5 - 7z^4 - 5\right)$

11. $-20t^4\left(3t^4 + 7t + 6\right)$

12. $c\left(-3c - 7\right)$

13. $3p\left(-p^4 + p^3 - 20\right)$

14. $-k^4\left(4k + 5\right)$

15. $-3\left(5m^4 - 7m + 9\right)$

16. $6x\left(-8x^3 + 20\right)$

17. $-w\left(w^4 - 5w + 8\right)$

18. $4y\left(7y^4 - y\right)$

19. $3d\left(d^7 - 8d^3 + 5\right)$

20. $-7t\left(-5t^4 - 9t + 2\right)$

21. $8\left(4w^4 - 20w + 5\right)$

22. $3y^4\left(y^4 - 22\right)$

23. $v^4\left(v^4 + 3v + 3\right)$

24. $9x\left(4x^3 + 3x + 2\right)$

25. $-7d\left(5d^4 + 8d - 4\right)$

26. $-k^4\left(-3k + 6\right)$

27. $3x\left(-x^4 - 7x + 7\right)$

28. $5z\left(5z^5 - z - 8\right)$

29. $-7y\left(20y^3 - 3\right)$

30. $4b^4\left(8b^4 + 5b + 5\right)$

4.8   Dividing Polynomials by Monomials

Example 11: $\dfrac{-8wx + 6x^2 - 16wx^2}{2wx}$

   Step 1:   Rewrite the problem. Divide each term from the top by the denominator, $2wx$.

   $$\dfrac{-8wx}{2wx} + \dfrac{6x^2}{2wx} + \dfrac{-16wx^2}{2wx}$$

   Step 2:   Simplify each term in the problem. Then combine like terms.

   $$-4 + \dfrac{3x}{w} - 8x$$

Simplify each of the following.

1. $\dfrac{bc^4 - 9bc - 4b^4c^4}{4bc}$

2. $\dfrac{3jk^4 + 24k + 20j^4k}{3jk}$

3. $\dfrac{7x^4y - 9xy^4 + 4y^3}{4xy}$

4. $\dfrac{26st^4 + st - 24s}{5st}$

5. $\dfrac{5wx^4 + 6wx - 24w^3}{4wx}$

6. $\dfrac{cd^4 + 20cd^3 + 26c^4}{4cd}$

7. $\dfrac{y^4z^3 - 4yz - 9z^4}{-4yz^4}$

8. $\dfrac{a^4b + 4ab^4 - 25ab^3}{4a^4}$

9. $\dfrac{pr^4 + 6pr + 9p^4r^4}{4pr^4}$

10. $\dfrac{6xy^4 - 3xy + 29x^4}{-3xy}$

11. $\dfrac{6x^4y + 24xy - 45y^4}{6xy}$

12. $\dfrac{7m^4n - 20mn - 47n^4}{7mn}$

13. $\dfrac{st^4 - 20st - 26s^4t^4}{4st}$

14. $\dfrac{8jk^4 - 25jk - 63j^4}{8jk}$

4.9 Removing Parentheses and Simplifying

In the following problem, you must multiply each term inside the parentheses by the numbers and variables outside the parentheses, and then add the polynomials to simplify the expressions.

Example 12: $9x\left(4x^4 - 7x + 8\right) - 3x\left(5x^4 + 3x - 9\right)$

Step 1:   Multiply to remove the first set of parentheses.

$$9x\left(4x^4 - 7x + 8\right) = 36x^5 - 63x^2 + 72x$$

Step 2:   Multiply to remove the second set of parentheses.

$$-3x\left(5x^4 + 3x - 9\right) = -15x^5 - 9x^2 + 27x$$

Step 3:   Copy each polynomial in columns, making sure the terms with the same variable and exponent are under each other. Add to simplify.

$$\begin{aligned} 36x^5 - 63x^2 + 72x \\ \underline{(+) - 15x^5 - 9x^2 + 27x} \\ 21x^5 - 72x^2 + 99x \end{aligned}$$

Remove the parentheses and simplify the following problems.

1. $5t\left(t + 8\right) + 7t\left(4t^4 - 5t + 2\right)$

2. $-7y\left(3y^4 - 7y + 3\right) - 6y\left(y^4 - 5y - 5\right)$

3. $-3\left(3x^4 + 5x\right) + 7x\left(x^4 + 3x + 4\right)$

4. $4b\left(7b^4 - 9b - 2\right) - 3b\left(5b + 3\right)$

5. $9d^4\left(3d + 5\right) - 8d\left(3d^4 + 5d + 7\right)$

6. $7a\left(3a^4 + 3a + 2\right) - \left(-4a^4 + 7a - 5\right)$

7. $3m\left(m + 8\right) + 9\left(5m^4 + m + 5\right)$

8. $5c^4\left(-6c^4 - 3c + 4\right) - 8c\left(7c^3 + 4c\right)$

9. $-9w\left(-w + 2\right) - 5w\left(3w - 7\right)$

10. $6p\left(4p^4 - 5p - 6\right) + 3p\left(p^4 + 6p + 20\right)$

4.10   Multiplying Two Binomials

When you multiply two binomials such as $(x + 6)(x - 7)$, you must multiply each term in the first binomial by each term in the second binomial. The easiest way is to use the **FOIL** method. If you can remember the word **FOIL**, it can help you keep order when you multiply. The "F" stands for **first**, "O" stands for **outside**, "I" stands for **inside**, and "L" stands for **last**.

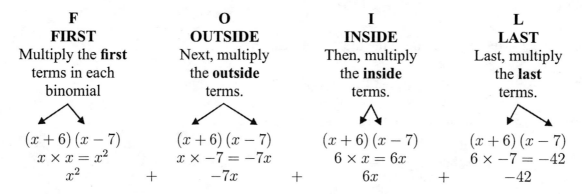

| **F** **FIRST** Multiply the **first** terms in each binomial | **O** **OUTSIDE** Next, multiply the **outside** terms. | **I** **INSIDE** Then, multiply the **inside** terms. | **L** **LAST** Last, multiply the **last** terms. |

$(x + 6)(x - 7)$ $(x + 6)(x - 7)$ $(x + 6)(x - 7)$ $(x + 6)(x - 7)$

$x \times x = x^2$ $x \times -7 = -7x$ $6 \times x = 6x$ $6 \times -7 = -42$

x^2 $+$ $-7x$ $+$ $6x$ $+$ -42

Now just combine like terms, $6x - 7x = -x$, and write your answer.
$(x + 6)(x - 7) = x^2 - x - 42$.

Note: It is customary for mathematicians to write polynomials in descending order. That means that the term with the highest-number exponent comes first in a polynomial. The next highest exponent is second and so on. When you use the **FOIL** method, the terms will always be in the customary order. You just need to combine like terms and write your answer.

Multiply the following binomials.

1. $(y - 8)(y + 3)$

2. $(4x + 5)(x + 20)$

3. $(5b - 3)(3b - 5)$

4. $(6g + 4)(g - 20)$

5. $(8k - 7)(-5k - 3)$

6. $(9v - 4)(3v + 5)$

7. $(20p + 4)(5p + 3)$

8. $(3h - 20)(-4h - 7)$

9. $(w - 5)(w - 8)$

10. $(6x + 2)(x - 4)$

11. $(7t + 3)(4t - 2)$

12. $(5y - 20)(5y + 20)$

13. $(a + 6)(3a + 7)$

14. $(3z - 9)(z - 5)$

15. $(7c + 4)(6c + 7)$

16. $(y + 3)(y - 3)$

17. $(4w - 7)(5w + 6)$

18. $(8x + 2)(x - 5)$

19. $(6t - 20)(5t - 5)$

20. $(7b + 6)(6b + 4)$

21. $(4z + 2)(20z + 5)$

22. $(22w - 9)(w + 3)$

23. $(7d - 20)(20d + 20)$

24. $(20g + 4)(g - 4)$

25. $(5p + 8)(4p + 3)$

27. $(9b - 9)(4b - 2)$

28. $(z + 3)(3z + 7)$

29. $(8y - 7)(y - 3)$

30. $(20x + 7)(3x - 2)$

31. $(3t + 2)(t + 20)$

32. $(4w - 20)(9w + 8)$

33. $(9s - 4)(s + 5)$

34. $(5k - 2)(9k + 20)$

35. $(h + 24)(h - 4)$

36. $(3x + 8)(8x + 3)$

37. $(4v - 6)(4v + 6)$

38. $(4x + 9)(4x - 3)$

39. $(k - 2)(6k + 24)$

40. $(3w + 22)(4w + 4)$

41. $(9y - 20)(7y - 3)$

42. $(6d + 23)(d - 2)$

43. $(8h + 3)(4h + 5)$

44. $(7n + 20)(7n - 7)$

45. $(6z + 7)(z - 9)$

46. $(5p + 7)(4p - 20)$

47. $(b + 4)(7b + 8)$

48. $(20y - 3)(9y - 8)$

26. $(m + 7)(m - 7)$

4.11   Factorials

A **factorial** is a mathematical operation represented by an exclamation point (!). This mathematical operation is only performed on positive integers. When it is applied, the integer is multiplied by every integer less than it, including 1. By definition, zero factorial is equal to 1, and 1 factorial is equal to 1. Using mathematical symbols, $0! = 1$, and $1! = 1$. Other factorials are best learned by example.

Examples: $5! = 5 \times 4 \times 3 \times 2 \times 1 = 120$
$10! = 10 \times 9 \times 8 \times 7 \times 6 \times 5 \times 4 \times 3 \times 2 \times 1 = 3,628,800$
$6! = 6 \times 5 \times 4 \times 3 \times 2 \times 1 = 720$
$3! = 3 \times 2 \times 1 = 6$
$1! = 1$
$0! = 1$

Factorials can also be used in mathematical operations such as multiplication and division. They are used to shorten the writing of some problems and as we will see later, and are used to solve problems involving permutations, combinations, and the binomial theorem.

Examples: $\dfrac{6!}{3!} = \dfrac{6 \times 5 \times 4 \times \cancel{3} \times \cancel{2} \times \cancel{1}}{\cancel{3} \times \cancel{2} \times \cancel{1}} = 6 \times 5 \times 4 = 120$

$5! \times 4! = (5 \times 4 \times 3 \times 2 \times 1) \times (4 \times 3 \times 2 \times 1) = (120) \times (24) = 2880$

Solve the following factorials.

1. $4!$

2. $7!$

3. $2!$

4. $8!$

5. $0!$

6. $11!$

7. $1!$

8. $\dfrac{5!}{3!}$

9. $3! \times 6!$

10. $7! \times 4!$

11. $\dfrac{8!}{5!}$

12. $0! \times 2!$

13. $\dfrac{10!}{0!}$

14. $4! \times 1!$

15. $\dfrac{8!}{3! \times 3!}$

16. $\dfrac{124!}{122!}$

17. $\dfrac{6!}{0! \times 4!}$

18. $\dfrac{12! \times 3!}{12!}$

4.12   The Combination Formula

The formula for combinations is used for counting problems, which we will see later in the book, and the binomial theorem in the next section. The formula is shown below.

$$\binom{n}{k} = \frac{n!}{k!\,(n-k)!}$$

The formula is read "n choose k."

Example 13:   Using the formula above, what is $\binom{n}{k}$ if $n = 5$ and $k = 2$?

Step 1:   Substitute that values for n and k into the formula.

$$\binom{n}{k} = \frac{n!}{k!\,(n-k)!} \rightarrow \binom{5}{2} = \frac{5!}{2!\,(5-2)!}$$

Step 2:   Solve.

$$\binom{5}{2} = \frac{5!}{2!\,(5-2)!} = \frac{5!}{2!3!} = \frac{5 \times 4 \times 3 \times 2 \times 1}{(2 \times 1)(3 \times 2 \times 1)} = \frac{5 \times 4}{2 \times 1} = \frac{20}{2} = 10$$

Using the combination formula, find the following.

1. $\binom{4}{0}$

2. $\binom{5}{1}$

3. $\binom{3}{0}$

4. $\binom{6}{1}$

5. $\binom{4}{1}$

6. $\binom{6}{2}$

7. $\binom{4}{2}$

8. $\binom{5}{3}$

9. $\binom{8}{3}$

10. $\binom{3}{1}$

4.13   Binomial Theorem

The purpose of the binomial theorem is to expand expressions raised to a power.

Example 14: $(x + y)^4 = (x + y)(x + y)(x + y)(x + y)$

It can be very awkward multiplying out a binomial raised to the 4th power, so we use the binomial theorem:

$$(x + y)^n = \sum_{k=0}^{n} \binom{n}{k} x^{n-k} y^k$$

Example 15:   Using the example $(x + y)^4$ we find that $n = 4$. So we will find the sum of the series from $k = 0$ to $k = 4$ and the pattern is $\binom{n}{k} x^{n-k} y^k$.

(Note: We will have 5 terms because the pattern goes from 0 to 4, which is 5 numbers.)

Step 1:   Starting with the 1st term, $k = 0$, we have

$$\binom{n}{k} x^{n-k} y^k = \binom{4}{0} x^{4-0} y^0 = \frac{4!}{0! \, (4 - 0)!} x^4 y^0 = \frac{4!}{1 \times 4!} x^4 y^0$$

$$= \frac{\cancel{4} \times \cancel{3} \times \cancel{2} \times \cancel{1}}{\cancel{4} \times \cancel{3} \times \cancel{2} \times \cancel{1}} x^4 y^0 = 1x^4 y^0 = x^4$$

Step 2:   For the 2nd term, $k = 1$, we have:

$$\binom{n}{k} x^{n-k} y^k = \binom{4}{1} x^{4-1} y^1 = \frac{4!}{1! \, (4 - 1)!} x^3 y^1 = \frac{4!}{1!3!} x^3 y^1$$

$$= \frac{4 \times \cancel{3} \times \cancel{2} \times \cancel{1}}{1 \times \cancel{3} \times \cancel{2} \times \cancel{1}} x^3 y = 4x^3 y$$

Step 3:   For the 3rd term, $k = 2$, we have:

$$\binom{n}{k} x^{n-k} y^k = \binom{4}{2} x^{4-2} y^2 = \frac{4!}{2! \, (4 - 2)!} x^2 y^2 = \frac{4!}{2!2!} x^2 y^2$$

$$= \frac{4 \times 3 \times \cancel{2} \times \cancel{1}}{2 \times 1 \times \cancel{2} \times \cancel{1}} x^2 y^2 = \frac{12}{2} x^2 y^2 = 6x^2 y^2$$

Step 4:   For the 4th term, $k = 3$, we have:

$$\binom{n}{k} x^{n-k}y^k = \binom{4}{3} x^{4-3}y^3 = \frac{4!}{3!\,(4-3)!}x^1y^3 = \frac{4!}{3!1!}x^1y^3$$

$$= \frac{4 \times 3 \times 2 \times 1}{3 \times 2 \times 1 \times 1}xy^3 = 4xy^3$$

Step 5:   For the 5th term, $k = 4$, we have:

$$\binom{n}{k} x^{n-k}y^k = \binom{4}{4} x^{4-4}y^4 = \frac{4!}{4!\,(4-4)!}x^0y^4 = \frac{4!}{4!0!}x^0y^4$$

$$= \frac{4!}{4! \times 1}x^0y^4 = \frac{4 \times 3 \times 2 \times 1}{4 \times 3 \times 2 \times 1}x^0y^4 = 1x^0y^4 = y^4$$

Step 6:   Now sum all of the terms together to expand the expression $(x + y)^4$.

$$(x + y)^4 = x^4 + 4x^3y + 6x^2y^2 + 4xy^3 + y^4$$

Use what you know about the Binomial Theorem to expand the binomials.

1. $(x + 4)^4$

2. $(x + 7)^2$

3. $(x - 3)^2$

4. $(x - 391)^1$

5. $(x + 2)^4$

6. $(x + 100)^0$

7. $(x - 1)^3$

8. $(x - 2)^4$

4.14   Perimeter and Area with Algebraic Expressions

You have already calculated the perimeter and area of various shapes with given measurements. You must also understand how to find the perimeter of shapes that are described by algebraic expressions. Study the examples below.

Example 16:   Use the equation $P = 2l + 2w$ to find the perimeter of the following rectangle.

Step 1:	Find $2l$.	$2(4x + 6) = 8x + 12$
Step 2:	Find $2w$.	$2(2x - 2) = 4x - 4$
Step 3:	Find $2l + 2w$.	$12x + 8$

Perimeter $= 12x + 8$

Example 17:   Using the formula $A = lw$, find the area of the rectangle below.

Step 1:	$A = (h - 2)(h + 1)$
Step 2:	$A = h^2 - 2h + h - 2$
Step 3:	$A = h^2 - h - 2$

Area $= h^2 - h - 2$

Example 18:   Find the area of the shaded part in the following figure.

Step 1:   Find the area of the larger rectangle.
$(4x + 6)(2x - 2) = 8x^2 - 8x + 12x - 12 = 8x^2 + 4x - 12$

Step 2:   Find the area of the smaller rectangle.
$(4x + 5)(x - 4) = 4x^2 - 16x + 5x - 20 = 4x^2 - 11x - 20$

Step 3:   Subtract the area of the smaller rectangle from the area of the larger rectangle.

$$
\begin{array}{ll}
8x^2 + 4x - 12 & 8x^2 + 4x - 12 \\
-(4x^2 - 11x - 20) & -4x^2 + 11x + 20 \quad \leftarrow \text{ Change signs} \\
\hline
& 4x^2 + 15x + 8 \quad \leftarrow \text{ Area of shaded section}
\end{array}
$$

Find the perimeter of each of the following rectangles.

1.

$6x - 4$
$2x + 8$

2.
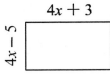
$4x + 3$
$4x - 5$

3.

$4x + 3$
$3x + 2$

4.

$7x - 6$
$5x + 7$

5.

$6x + 3$
$4x + 2$

6.
$2x + 4$
$x - 2$

Find the area of each of the following rectangles.

7.

$5 - 2m$
$2 + m$

8.
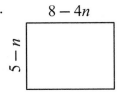
$8 - 4n$
$5 - n$

9.
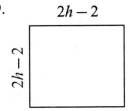
$2h - 2$
$2h - 2$

10.
$9 - h$
$4 + 2h$

11.
n
$n + 8$

12.
$7 + 2b$
$2 + b$

Find the area of the shaded portion of each figure below.

13.

$8x - 7$
$7x + 2$
$6x + 3$
$2x + 2$

14.

$4x - 1$
$3x - 4$
$3x + 2$
$2x - 1$

15.

$6x + 3$
$3x - 4$
$5x + 4$
$5x - 5$

16.
$11x + 3$
$9x - 4$
$10x + 8$
$3x + 2$

17.

$9x - 5$
$5x - 7$
$3x + 4$
$x + 5$

18.
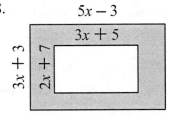
$5x - 3$
$3x + 5$
$3x + 3$
$2x + 7$

4.15   Volume with Algebraic Expressions

Volume is the amount of space that can be filled by an object expressed in cubic units.

Equations: **Cube**   **Rectangular Prism**

$$V = s \times s \times s = s^3$$ $$V = l \times w \times h$$

    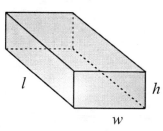

Example 19:   Find the volume of a cube with side length $(x - 2)$.

Step 1:   Write the equation: $V = (x - 2)^3 = (x - 2) \times (x - 2) \times (x - 2)$

Step 2:   FOIL the first two factors: $V = (x^2 - 4x - 4)(x - 2)$

Step 3:   FOIL again: $V = x^3 - 4x^2 - 4x - 2x^2 + 8x + 8$

Step 4:   Simplify: $V = x^3 - 6x^2 + 4x + 8$

Example 20:   Find the volume of a rectangular prism with height x, length $(x + 1)$, and width $(x - 7)$.

Step 1:   Substitute values into equation: $V = l \times w \times h = (x + 1)(x - 7)(x)$

Step 2:   FOIL the first two factors: $V = (x^2 - 7x + x - 7)(x)$

Step 3:   Distribute: $V = x^3 - 7x^2 + x^2 - 7x$

Step 4:   Simplify: $V = x^3 - 6x^2 - 7x$

Find the volume of the figure.

1. A cube with side length $(x + 7)$

2. A cube with side length $(3 + x)$

3. A cube with side length $(x - 9)$

4. Find the volume of the cube in number 1 if $x = 2$

5. A rectangular prism with length $(x + 2)$, width $(x + 3)$, and height (x)

6. A rectangular prism with width $(x - 2)$, length $(7 + x)$, height $(x + 9)$

7. A rectangular prism with length $(x + 1)$, height $(x + 4)$, and width $(12 + x)$

8. Find the volume of number 6 if $x = 7$

Chapter 4 Review

Simplify.

1. $3a^4 + 20a^4$

2. $(8x^4y^5)(20xy^7)$

3. $-6z^4(z+3)$

4. $(5b^4)(7b^3)$

5. $8x^4 - 20x^4$

6. $(7p-5)-(3p+4)$

7. $-7t(3t+20)^2$

8. $(3w^3y^4)(5wy^7)$

9. $3(4g+3)^2$

10. $25d^5 - 20d^5$

11. $(8w-5)(w-9)$

12. $27t^4 + 5t^4$

13. $(8c^5)(20c^4)$

14. $(20x+4)(x+7)$

15. $5y(5y^4 - 20y + 4)$

16. $(9a^5b)(4ab^3)(ab)$

17. $(7w^6)(20w^{20})$

18. $9x^3 + 24x^3$

19. $27p^7 - 22p^7$

20. $(3s^5t^4)(5st^3)$

21. $(5d+20)(4d+8)$

22. $5w(-3w^4 + 8w - 7)$

23. $45z^6 - 20z^6$

24. $-8y^3 - 9y^3$

25. $(8x^5)(8x^7)$

26. $28p^4 + 20p^4$

27. $(a^4v)(4av)(a^3v^6)$

28. $5(6y-7)^2$

29. $(3c^4)(6c^9)$

30. $(5x^7y^3)(4xy^3)$

31. Add $4x^4 + 20x$ and $7x^4 - 9x + 4$

32. $5t(6t^4 + 5t - 6) + 9t(3t+3)$

33. Subtract $y^4 + 5y - 6$ from $3y^4 + 8$

34. $4x(5x^4 + 6x - 3) + 5x(x+3)$

35. $(6t-5)-(6t^4 + t - 4)$

36. $(5x+6)+(8x^4 - 4x + 3)$

37. Subtract $7a - 4$ from $a + 20$

38. $(-4y+5)+(5y-6)$

39. $4t(t+6)-7t(4t+8)$

40. Add $3c - 5$ and $c^4 - 3c - 4$

41. $4b(b-5)-(b^4 + 4b + 2)$

42. $(6k^4 + 7k)+(k^4 + k + 20)$

43. $(q^4r^3)(3qr^4)(4q^5r)$

44. $(7df)(d^5f^4)(4df)$

45. $(8g^4h^3)(g^3h^6)(6gh^3)$

46. $(9v^4x^3)(3v^6x^4)(4v^5x^5)$

47. $(3n^4m^4)(20n^4m)(n^3m^8)$

48. $(22t^4a^4)(5t^3a^9)(4t^6a)$

49. $\dfrac{24(4a^3)b}{3a^4b^{-4}}$

50. $\dfrac{8(g^3h^3)}{5(g^4h)^{-4}}$

51. $\dfrac{26(m^4n^3)^4}{5(m^4n)^{-4}}$

52. $\dfrac{25p^3q^3}{4p^4q}$

53. $\dfrac{9(e^5h^{-4})^{-4}}{36e^4h^7}$

54. $\dfrac{44x^3y^5}{154(x^{-3}y^8)^4}$

Use what you know about volume to solve the following questions.

55. Find the volume of a rectangular prism with length $(x-6)$, width $(9+x)$, and height $(x-11)$.

56. Find the volume of the rectangular prism in question 55 when $x = 3$.

57. Find the volume of a cube with side length $(x+4)$.

58. Find the volume of the cube in question 57 when $x = 4$.

Use what you know about the Binomial Theorem to expand the following binomials.

59. $(x+6)^0 =$

60. $(x+3)^1 =$

61. $(x+5)^2 =$

62. $(x+1)^3 =$

63. $(x+2)^4 =$

64. $(x+1)^5 =$

Chapter 4 Test

1. $2x^2 + 5x^2 =$

   (A) $10x^4$
   (B) $7x^4$
   (C) $7x^2$
   (D) $10x^2$

2. $-8m^3 + m^3 =$

   (A) $-8m^6$
   (B) $-8m^9$
   (C) $-9m^6$
   (D) $-7m^3$

3. $(6x^3 + x^2 - 5) + (-3x^3 - 2x^2 + 4) =$

   (A) $3x^3 - x^2 - 1$
   (B) $3x^3 - 3x^2 - 1$
   (C) $3x^3 - 3x^2 - 9$
   (D) $-3x^3 - 3x^2 - 1$

4. $(-7c^2 + 5c + 3) + (-c^2 - 7c + 2) =$

   (A) $-3x^3 - 3x^2 - 1$
   (B) $-8c^2 - 2c + 5$
   (C) $-6c^2 - 12c + 5$
   (D) $-8c^2 - 12c + 5$

5. $(5x^3 - 4x^2 + 5) - (-2x^3 - 3x^2) =$

   (A) $3x^3 + x^2 + 5$
   (B) $3x^3 - 7x^2 + 5$
   (C) $7x^3 - x^2 + 5$
   (D) $7x^3 - 7x^2 + 5$

6. $(-z^3 - 4z^2 - 6) - (3z^3 - 6z + 5) =$

   (A) $-4z^3 - 4z^2 + 6z - 11$
   (B) $-2z^3 - 10z - 1$
   (C) $-4z^3 - 10z^2 - 1$
   (D) $-2z^2 + 2z - 11$

7. $(-7d^5)(-3d^2) =$

   (A) $-21d^7$
   (B) $21d^{10}$
   (C) $21d^7$
   (D) $-21d^{10}$

8. $(-5c^3d)(3c^5d^3)(2cd^4) =$

   (A) $30c^{15}d^8$
   (B) $15c^8d^{12}$
   (C) $-17c^{15}d^{12}$
   (D) $-30c^9d^8$

9. $-11j^2 \times -j^4 =$

   (A) $11j^6$
   (B) $11j^8$
   (C) $-11j^6$
   (D) $-11j^8$

10. $-6m^2(7m^2 + 5m - 6) =$

    (A) $-42m^2 + 30m^3 - 36$
    (B) $-42m^4 - 30m^3 + 36m^2$
    (C) $-13m^4 - m^2 + 36m^2$
    (D) $42m^4 - 30m^3 - 36m^2$

11. $-h^2(-4h + 5) =$

    (A) $-4h^3 - 5h^2$
    (B) $4h^3 - 5h^2$
    (C) $-5h^2 - 5h^2$
    (D) $-5h^3 - 5h^2$

12. $\dfrac{4xy^2 - 6xy + 8x^2y}{2xy} =$

    (A) $2xy - 3 + 4x$
    (B) $2y - 3 + 4xy$
    (C) $2y - 3 + 4x$
    (D) $2xy - 3 + 4x^2$

13. $\dfrac{3cd^3 + 6c^2d - 12cd}{3cd} =$

  (A) $cd + 3c - 4cd$
  (B) $d^2 + 2c - 4cd$
  (C) $cd^2 + 2c - 4cd$
  (D) $d^2 + 2c - 4$

14. $4m(m - 5) + 3m(2m^2 - 6m + 4) =$

  (A) $6m^3 - 14m^2 - 8m$
  (B) $-8m^2 - 8m - 1$
  (C) $7m - 14m^2 - 1$
  (D) $10m^2 - 26m - 20$

15. $2h(3h^2 - 5h - 2) + 4h(h^2 + 6h + 8) =$

  (A) $6h^3 + 19h^2 + 28h$
  (B) $-8m^2 - 8m - 1$
  (C) $7m - 14m^2 - 1$
  (D) $10h^3 + 14h^2 + 28h$

16. Multiply the following binomial and simplify. $(x - 3)(x + 3)$

  (A) $x^2 - 3x + 3x - 9$
  (B) $x^2 - 9$
  (C) $x^2 + 9$
  (D) $x^2 + 6x + 9$

17. Multiply the following binomial and simplify. $(x + 9)(x + 1)$

  (A) $x^2 + 10x + 9$
  (B) $x^2 + 10x + 10$
  (C) $x^2 + 9x + 9$
  (D) $x^2 + 9x + x + 9$

18. Multiply the following binomial and simplify. $(x - 2)^2$

  (A) $x^2 - 4x - 4$
  (B) $x^2 - 2x + 4$
  (C) $x^2 - 2x - 4$
  (D) $x^2 - 4x + 4$

19. The length of the rectangle is 5 units longer than the width. Which expression could be used to represent the area of the rectangle?

  (A) $w^2 + 5w$
  (B) $w^2 + 5$
  (C) $w^2 + 25$
  (D) $w^2 + 10w + 25$

20. The volume of a cube with side length $(x - 3)$ is 8, what is x?

  (A) 3
  (B) 4
  (C) 5
  (D) 11

21. Given the figure with length $(x - 4)$, width $(x - 2)$, and height (x), what is the volume when $x = 7$?

  (A) 100
  (B) 105
  (C) 130
  (D) 115

22. $(x + 4)^2 = ?$

  (A) $x^2 + 4$
  (B) $x^2 + 16$
  (C) $x^2 + 16x + 8$
  (D) $x^2 + 8x + 16$

23. How many terms are in the expansion of $(x + 4)^4$?

  (A) 4
  (B) 5
  (C) 6
  (D) 3

24. When expanding $(x + 71)^5$, the sum of the exponents of each term must equal _____ ?

  (A) 4
  (B) 5
  (C) 6
  (D) 7

Chapter 5
Factoring

This chapter covers the following Georgia Performance Standards:

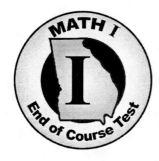

MM1A	Algebra	MM1A2e
		MM1A2f

5.1 Finding the Greatest Common Factor of Polynomials

In a multiplication problem, the numbers multiplied together are called **factors**. The answer to a multiplication problem is a called the **product**.

In the multiplication problem $5 \times 4 = 20$, 5 and 4 are factors and 20 is the product.

If we reverse the problem, $20 = 5 \times 4$, we say we have **factored** 20 into 5×4.

In this chapter, we will factor **polynomials**.

Example 1:   Find the greatest common factor of $2y^3 + 6y^2$.

   Step 1:   Look at the whole numbers. The greatest common factor of 2 and 6 is 2. Factor the 2 out of each term.

   $2\left(y^3 + 3y^2\right)$

   Step 2:   Look at the remaining terms, $y^3 + 3y^2$. What are the common factors of each term?

$$
\begin{array}{rcl}
y^3 & = & y \times \boxed{y \times y} \\
3y^2 & = & 3 \times \boxed{y \times y}
\end{array}
\longleftarrow \text{ common factors } = y^2
$$

   Step 3:   Factor 2 and y^2 out of each term: $2y^2\left(y + 3\right)$

   Check: $2y^2\left(y + 3\right) = 2y^3 + 6y^2$

Factor by finding the greatest common factor in each of the following.

1. $6x^4 + 18x^2$

2. $14y^3 + 7y$

3. $4b^5 + 12b^3$

4. $10a^3 + 5$

5. $2y^3 + 8y^2$

6. $6x^4 - 12x^2$

7. $18y^2 - 12y$

8. $15a^3 - 25a^2$

9. $4x^3 + 16x^2$

10. $6b^2 + 21b^5$

11. $27m^3 + 18m^4$

12. $100x^4 - 25x^3$

13. $4b^4 - 12b^3$

14. $18c^2 + 24c$

15. $20y^3 + 30y^5$

16. $16x^2 - 24x^5$

17. $15a^4 - 25a^2$

18. $24b^3 + 16b^6$

19. $36y^4 + 9y^2$

20. $42x^3 + 49x$

Factoring larger polynomials with 3 or 4 terms works the same way.

Example 2: $4x^5 + 16x^4 + 12x^3 + 8x^2$

Step 1: Find the greatest common factor of the whole numbers. 4 can be divided evenly into 4, 16, 12, and 8; therefore, 4 is the greatest common factor.

Step 2: Find the greatest common factor of the variables. x^5, x^4, x^3, and x^2 can be divided by x^2, the lowest power of x in each term.

$$4x^5 + 16x^4 + 12x^3 + 8x^2 = 4x^2 \left(x^3 + 4x^2 + 3x + 2 \right)$$

Factor each of the following polynomials.

1. $5a^3 + 15a^2 + 20a$

2. $18y^4 + 6y^3 + 24y^2$

3. $12x^5 + 21x^3 + x^2$

4. $6b^4 + 3b^3 + 15b^2$

5. $14c^3 + 28c^2 + 7c$

6. $15b^4 - 5b^2 + 20b$

7. $t^3 + 3t^2 - 5t$

8. $8a^3 - 4a^2 + 12a$

9. $16b^5 - 12b^4 - 10b^2$

10. $20x^4 + 16x^3 - 24x^2 + 28x$

11. $40b^7 + 30b^5 - 50b^3$

12. $20y^4 - 15y^3 + 30y^2$

13. $4m^5 + 8m^4 + 12m^3 + 6m^2$

14. $16x^5 + 20x^4 - 12x^3 + 24x^2$

15. $18y^4 + 21y^3 - 9y^2$

16. $3n^5 + 9n^3 + 12n^2 + 15n$

17. $4d^6 - 8d^2 + 2d$

18. $10w^2 + 4w + 2$

19. $6t^3 - 3t^2 + 9t$

20. $25p^5 - 10p^3 - 5p^2$

21. $18x^4 + 9x^2 - 36x$

22. $6b^4 - 12b^2 - 6b$

23. $y^3 + 3y^2 - 9y$

24. $10x^5 - 2x^4 + 4x^2$

Example 3:   Find the greatest common factor of $4a^3b^2 - 6a^2b^2 + 2a^4b^3$

   Step 1:   The greatest common factor of the whole numbers is 2.

$$4a^3b^2 - 6a^2b^2 + 2a^4b^3 = 2\left(2a^3b^2 - 3a^2b^2 + a^4b^3\right)$$

   Step 2:   Find the lowest power of each variable that is in each term. Factor them out of each term. The lowest power of a is a^2. The lowest power of b is b^2.

$$4a^3b^2 - 6a^2b^2 + 2a^4b^3 = 2a^2b^2\left(2a - 3 + a^2b\right)$$

Factor each of the following polynomials.

1. $3a^2b^2 - 6a^3b^4 + 9a^2b^3$

2. $12x^4y^3 + 18x^3y^4 - 24x^3y^3$

3. $20x^2y - 25x^3y^3$

4. $12x^2y - 20x^2y^2 + 16xy^2$

5. $8a^3b + 12a^2b + 20a^2b^3$

6. $36c^4 + 42c^3 + 24c^2 - 18c$

7. $14m^3n^4 - 28m^3n^2 + 42m^2n^3$

8. $16x^4y^2 - 24x^3y^2 + 12x^2y^2 - 8xy^2$

9. $32c^3d^4 - 56c^2d^3 + 64c^3d^2$

10. $21a^4b^3 + 27a^2b^3 + 15a^3b^2$

11. $4w^3t^2 + 6w^2t - 8wt^2$

12. $5pq^3 - 2p^2q^2 - 9p^3q$

13. $49x^3t^3 + 7xt^2 - 14xt^3$

14. $9cd^4 - 3d^4 - 6c^2d^3$

15. $12a^2b^3 - 14ab + 10ab^2$

16. $25x^4 + 10x - 20x^2$

17. $bx^3 - b^2x^2 + b^3x$

18. $4k^3a^2 + 22ka + 16k^2a^2$

19. $33w^4y^2 - 9w^3y^2 + 24w^2y^2$

20. $18x^3 - 9x^5 + 27x^2$

5.2 Factor By Grouping

Not all polynomials have a common factor in each term. In this case they may sometimes be factored by grouping.

Example 4:   Factor $ab + 4a + 2b + 8$

Step 1:   Factor an a from the first two terms and a 2 from the last two terms.

$$a(b + 4) + 2(b + 4)$$

Now the polynomial has two terms, $a(b + 4)$ and $2(b + 4)$. Notice that $(b + 4)$ is a factor of each term.

Step 2:   Factor out the common factor of each term:

$$ab + 4a + 2b + 8 = (b + 4)(a + 2).$$

Check: Multiply using the FOIL method to check.

$$(b + 4)(a + 2) = ab + 4a + 2b + 8$$

Factor the following polynomials by grouping.

1. $xy + 4x + 2y + 8$

2. $cd + 5c + 4d + 20$

3. $xy - 4x + 6y - 24$

4. $ab + 6a + 3b + 18$

5. $ab + 3a - 5b - 15$

6. $xy - 2x + 6x - 12$

7. $cd + 4c + 4d + 16$

8. $mn - 5m + 3n - 15$

9. $ab + 4a + 3b + 12$

10. $xy + 7x - 4y - 28$

11. $ab - 2a + 8b - 16$

12. $cd + 4c - 5d - 20$

13. $mn + 6m - 2n - 12$

14. $xy - 9x - 3y + 27$

15. $bc - 3b + 5c - 15$

16. $ab + a + 7b + 7$

17. $xy + 4y + 2y + 8$

18. $cd + 9c - d - 9$

19. $ab + 2a - 7b - 14$

20. $xy - 6x - 2y + 12$

21. $wz + 6z - 4w - 24$

5.3   Finding the Numbers

The next kind of factoring we will do requires thinking of two numbers with a certain sum and a certain product.

Example 5:   Which two numbers have a sum of 8 and a product of 12? In other words, what pair of numbers would answer both equations?

$$\underline{\hspace{1cm}} + \underline{\hspace{1cm}} = 8 \quad \text{and} \quad \underline{\hspace{1cm}} \times \underline{\hspace{1cm}} = 12$$

You may think $4 + 4 = 8$, but 4×4 does not equal 12.
Or you may think $7 + 1 = 8$, but 7×1 does not equal 12.

$6 + 2 = 8$ and $6 \times 2 = 12$, so 6 and 2 are the pair of numbers that will work in both equations.

For each problem below, find one pair of numbers that will solve both equations.

1. $\underline{\hspace{1cm}} + \underline{\hspace{1cm}} = 14$ and $\underline{\hspace{1cm}} \times \underline{\hspace{1cm}} = 40$

2. $\underline{\hspace{1cm}} + \underline{\hspace{1cm}} = 10$ and $\underline{\hspace{1cm}} \times \underline{\hspace{1cm}} = 21$

3. $\underline{\hspace{1cm}} + \underline{\hspace{1cm}} = 18$ and $\underline{\hspace{1cm}} \times \underline{\hspace{1cm}} = 81$

4. $\underline{\hspace{1cm}} + \underline{\hspace{1cm}} = 12$ and $\underline{\hspace{1cm}} \times \underline{\hspace{1cm}} = 20$

5. $\underline{\hspace{1cm}} + \underline{\hspace{1cm}} = 7$ and $\underline{\hspace{1cm}} \times \underline{\hspace{1cm}} = 12$

6. $\underline{\hspace{1cm}} + \underline{\hspace{1cm}} = 8$ and $\underline{\hspace{1cm}} \times \underline{\hspace{1cm}} = 15$

7. $\underline{\hspace{1cm}} + \underline{\hspace{1cm}} = 10$ and $\underline{\hspace{1cm}} \times \underline{\hspace{1cm}} = 25$

8. $\underline{\hspace{1cm}} + \underline{\hspace{1cm}} = 14$ and $\underline{\hspace{1cm}} \times \underline{\hspace{1cm}} = 48$

9. $\underline{\hspace{1cm}} + \underline{\hspace{1cm}} = 12$ and $\underline{\hspace{1cm}} \times \underline{\hspace{1cm}} = 36$

10. $\underline{\hspace{1cm}} + \underline{\hspace{1cm}} = 17$ and $\underline{\hspace{1cm}} \times \underline{\hspace{1cm}} = 72$

11. $\underline{\hspace{1cm}} + \underline{\hspace{1cm}} = 15$ and $\underline{\hspace{1cm}} \times \underline{\hspace{1cm}} = 56$

12. $\underline{\hspace{1cm}} + \underline{\hspace{1cm}} = 9$ and $\underline{\hspace{1cm}} \times \underline{\hspace{1cm}} = 18$

13. $\underline{\hspace{1cm}} + \underline{\hspace{1cm}} = 13$ and $\underline{\hspace{1cm}} \times \underline{\hspace{1cm}} = 40$

14. $\underline{\hspace{1cm}} + \underline{\hspace{1cm}} = 16$ and $\underline{\hspace{1cm}} \times \underline{\hspace{1cm}} = 63$

15. $\underline{\hspace{1cm}} + \underline{\hspace{1cm}} = 10$ and $\underline{\hspace{1cm}} \times \underline{\hspace{1cm}} = 16$

16. $\underline{\hspace{1cm}} + \underline{\hspace{1cm}} = 8$ and $\underline{\hspace{1cm}} \times \underline{\hspace{1cm}} = 16$

17. $\underline{\hspace{1cm}} + \underline{\hspace{1cm}} = 9$ and $\underline{\hspace{1cm}} \times \underline{\hspace{1cm}} = 20$

18. $\underline{\hspace{1cm}} + \underline{\hspace{1cm}} = 13$ and $\underline{\hspace{1cm}} \times \underline{\hspace{1cm}} = 36$

19. $\underline{\hspace{1cm}} + \underline{\hspace{1cm}} = 15$ and $\underline{\hspace{1cm}} \times \underline{\hspace{1cm}} = 50$

20. $\underline{\hspace{1cm}} + \underline{\hspace{1cm}} = 11$ and $\underline{\hspace{1cm}} \times \underline{\hspace{1cm}} = 30$

5.4 More Finding the Numbers

Now that you have mastered positive numbers, take up the challenge of finding pairs of negative numbers or pairs where one number is negative and one is positive.

Example 6: Which two numbers have a sum of -3 and a product of -40? In other words, what pair of numbers would answer both equations?

$$\underline{\hspace{1cm}} + \underline{\hspace{1cm}} = -3 \quad \text{and} \quad \underline{\hspace{1cm}} \times \underline{\hspace{1cm}} = -40$$

It is faster to look at the factors of 40 first. 8 and 5 and 10 and 4 are possibilities. 8 and 5 have a difference of 3, and in fact, $5 + (-8) = -3$ and $5 \times (-8) = -40$. This pair of numbers, 5 and -8, will satisfy both equations.

For each problem below, find one pair of numbers that will solve both equations.

1. $\underline{\hspace{1cm}} + \underline{\hspace{1cm}} = -2$ and $\underline{\hspace{1cm}} \times \underline{\hspace{1cm}} = -35$
2. $\underline{\hspace{1cm}} + \underline{\hspace{1cm}} = 4$ and $\underline{\hspace{1cm}} \times \underline{\hspace{1cm}} = -5$
3. $\underline{\hspace{1cm}} + \underline{\hspace{1cm}} = 4$ and $\underline{\hspace{1cm}} \times \underline{\hspace{1cm}} = -12$
4. $\underline{\hspace{1cm}} + \underline{\hspace{1cm}} = -6$ and $\underline{\hspace{1cm}} \times \underline{\hspace{1cm}} = 8$
5. $\underline{\hspace{1cm}} + \underline{\hspace{1cm}} = 3$ and $\underline{\hspace{1cm}} \times \underline{\hspace{1cm}} = -40$
6. $\underline{\hspace{1cm}} + \underline{\hspace{1cm}} = 10$ and $\underline{\hspace{1cm}} \times \underline{\hspace{1cm}} = -11$
7. $\underline{\hspace{1cm}} + \underline{\hspace{1cm}} = 6$ and $\underline{\hspace{1cm}} \times \underline{\hspace{1cm}} = -27$
8. $\underline{\hspace{1cm}} + \underline{\hspace{1cm}} = 8$ and $\underline{\hspace{1cm}} \times \underline{\hspace{1cm}} = -20$
9. $\underline{\hspace{1cm}} + \underline{\hspace{1cm}} = -5$ and $\underline{\hspace{1cm}} \times \underline{\hspace{1cm}} = -24$
10. $\underline{\hspace{1cm}} + \underline{\hspace{1cm}} = -3$ and $\underline{\hspace{1cm}} \times \underline{\hspace{1cm}} = -28$
11. $\underline{\hspace{1cm}} + \underline{\hspace{1cm}} = -2$ and $\underline{\hspace{1cm}} \times \underline{\hspace{1cm}} = -48$
12. $\underline{\hspace{1cm}} + \underline{\hspace{1cm}} = -1$ and $\underline{\hspace{1cm}} \times \underline{\hspace{1cm}} = -20$
13. $\underline{\hspace{1cm}} + \underline{\hspace{1cm}} = -3$ and $\underline{\hspace{1cm}} \times \underline{\hspace{1cm}} = 2$
14. $\underline{\hspace{1cm}} + \underline{\hspace{1cm}} = 1$ and $\underline{\hspace{1cm}} \times \underline{\hspace{1cm}} = -30$
15. $\underline{\hspace{1cm}} + \underline{\hspace{1cm}} = -7$ and $\underline{\hspace{1cm}} \times \underline{\hspace{1cm}} = 12$
16. $\underline{\hspace{1cm}} + \underline{\hspace{1cm}} = 6$ and $\underline{\hspace{1cm}} \times \underline{\hspace{1cm}} = -16$
17. $\underline{\hspace{1cm}} + \underline{\hspace{1cm}} = 5$ and $\underline{\hspace{1cm}} \times \underline{\hspace{1cm}} = -24$
18. $\underline{\hspace{1cm}} + \underline{\hspace{1cm}} = -4$ and $\underline{\hspace{1cm}} \times \underline{\hspace{1cm}} = 4$
19. $\underline{\hspace{1cm}} + \underline{\hspace{1cm}} = -1$ and $\underline{\hspace{1cm}} \times \underline{\hspace{1cm}} = -42$
20. $\underline{\hspace{1cm}} + \underline{\hspace{1cm}} = -6$ and $\underline{\hspace{1cm}} \times \underline{\hspace{1cm}} = 8$

5.5 Factoring Trinomials

In the chapter on polynomials, you multiplied binomials (two terms) together, and the answer was a trinomial (three terms).

For example, $(x + 6)(x - 5) = x^2 + x - 30$

Now, you need to practice factoring a trinomial into two binomials.

Example 7:   Factor $x^2 + 6x + 8$

     Step 1:   When the trinomial is in descending order as in the example above, you need to find a pair of numbers whose sum equals the number in the second term, while their product equals the third term. In the above example, find the pair of numbers that has a sum of 6 and a product of 8.

$$\underline{\hspace{1cm}} + \underline{\hspace{1cm}} = 6 \quad \text{and} \quad \underline{\hspace{1cm}} \times \underline{\hspace{1cm}} = 8$$

                The pair of numbers that satisfy both equations is 4 and 2.

     Step 2:   Use the pair of numbers in the binomials.

                The factors of $x^2 + 6x + 8$ are $(x + 4)(x + 2)$

     Check: To check, use the FOIL method.
                $(x + 4)(x + 2) = x^2 + 4x + 2x + 8 = x^2 + 6x + 8$

                Notice, when the second term and the third term of the trinomial are both positive, both numbers in the solution are positive.

Example 8:   Factor $x^2 - x - 6$ Find the pair of numbers where:

the sum is -1 and the product is -6

$$\underline{\hspace{1cm}} + \underline{\hspace{1cm}} = -1 \quad \text{and} \quad \underline{\hspace{1cm}} \times \underline{\hspace{1cm}} = -6$$

The pair of numbers that satisfies both equations is 2 and -3.
The factors of $x^2 - x - 6$ are $(x + 2)(x - 3)$

Notice, if the second term and the third term are negative, one number in the solution pair is positive, and the other number is negative.

Example 9:   Factor $x^2 - 7x + 12$ Find the pair of numbers where:

the sum is -7 and the product is 12

_____ + _____ = -7 and   _____ × _____ = 12

The pair of numbers that satisfies both equations is -3 and -4
The factors of $x^2 - 7x + 12$ are $(x - 3)(x - 4)$.

Notice, if the second term of a trinomial is negative and the third term is positive, both numbers in the solution are negative.

Find the factors of the following trinomials.

1. $x^2 - x - 2$

2. $y^2 + y - 6$

3. $w^2 + 3w - 4$

4. $t^2 + 5t + 6$

5. $x^2 + 2x - 8$

6. $k^2 - 4k + 3$

7. $t^2 + 3t - 10$

8. $x^2 - 3x - 4$

9. $y^2 - 5y + 6$

10. $y^2 + y - 20$

11. $a^2 - a - 6$

12. $b^2 - 4b - 5$

13. $c^2 - 5c - 14$

14. $c^2 - c - 12$

15. $d^2 + d - 6$

16. $x^2 - 3x - 28$

17. $y^2 + 3y - 18$

18. $a^2 - 9a + 20$

19. $b^2 - 2b - 15$

20. $c^2 + 7c - 8$

21. $t^2 - 11t + 30$

22. $w^2 + 13w + 36$

23. $m^2 - 2m - 48$

24. $y^2 + 14y + 49$

25. $x^2 + 7x + 10$

26. $a^2 - 7a + 6$

27. $d^2 - 6d - 27$

5.6   Factoring More Trinomials

Some trinomials have a whole number in front of the first term that cannot be factored out of the trinomial. The trinomial can still be factored.

Example 10:   Factor $2x^2 + 5x - 3$

Step 1:   To get a product of $2x^2$, one factor must begin with $2x$ and the other with x.

$(2x \quad)(x \quad)$

Step 2:   Now think: What two numbers give a product of -3? The two possibilities are 3 and -1 or -3 and 1. We know they could be in any order so there are 4 possible arrangements.

$(2x + 3)(x - 1)$
$(2x - 3)(x + 1)$
$(2x + 1)(x - 3)$
$(2x - 1)(x + 3)$

Step 3:   Multiply each possible answer until you find the arrangement of the numbers that works. Multiply the outside terms and the inside terms and add them together to see which one will equal $5x$.

$(2x + 3)(x - 1) = 2x^2 + x - 3$
$(2x - 3)(x + 1) = 2x^2 - x - 3$
$(2x + 1)(x - 3) = 2x^2 - 5 - 3$
$\boxed{(2x - 1)(x + 3) = 2x^2 + 5x - 3}$ ← This arrangement works, therefore:

The factors of $2x^2 + 5x - 3$ are $(2x - 1)(x + 3)$

Alternative: You can do some of the multiplying in your head. For the above example, ask yourself the following question: What two numbers give a product of -3 and give a sum of 5 (the whole number in the second term) when one number is first multiplied by 2 (the whole number in front of the first term)? The pair of numbers, -1 and 3, have a product of -3 and a sum of 5 when the 3 is first multiplied by 2. Therefore, the 3 will go in the opposite factor of the $2x$ so that when the terms are multiplied, you get -5.

You can use this method to at least narrow down the possible pairs of numbers when you have several from which to choose.

Factor the following trinomials.

1. $3y^2 + 14y + 8$

2. $5a^2 + 24a - 5$

3. $7b^2 + 30b + 8$

4. $2c^2 - 9c + 9$

5. $2y^2 - 7y - 15$

6. $3x^2 + 4x + 1$

7. $7y^2 + 13y - 2$

8. $11a^2 + 35a + 6$

9. $5y^2 + 17y - 12$

10. $3a^2 + 4a - 7$

11. $2a^2 + 3a - 20$

12. $5b^2 - 13b - 6$

13. $3y^2 - 17x + 36$

14. $2x^2 - 17x + 36$

15. $11x^2 - 29x - 12$

16. $5c^2 + 2c - 16$

17. $7y^2 - 30y + 27$

18. $2x^2 - 3x - 20$

19. $5b^2 + 24b - 5$

20. $7d^2 + 18d + 8$

21. $3x^2 - 20x + 25$

22. $2a^2 - 7a - 4$

23. $5m^2 + 12m + 4$

24. $9y^2 - 5y - 4$

25. $2b^2 - 13b + 18$

26. $7x^2 + 31x - 20$

27. $3c^2 - 2c - 21$

5.7   Factoring Trinomials with Two Variables

Some trinomials have two variables with exponents. You can still factor these trinomials.

Example 11:   Factor $x^2 + 5xy + 6y^2$

Step 1:   Notice there is an x^2 in the first term and a y^2 in the last term. When you see two different terms that are squared, you know there has to be an x and a y in each factor:

$$(x \quad y)(x \quad y)$$

Step 2:   Now think: What are two numbers whose sum is 5 and product is 6? You see that 3 and 2 will work. Put 3 and 2 in the factors:

$$(x + 3y)(x + 2y)$$

Check: Multiply to check. $(x + 3y)(x + 2y) = x^2 + 3xy + 2xy + 6y^2 = x^2 + 5xy + 6y^2$

Factor the following trinomials.

1. $a^2 + 6ab + 8b^2$

2. $x^2 + 3xy - 4y^2$

3. $c^2 - 2cd - 15d^2$

4. $g^2 + 7gh + 10h^2$

5. $a^2 - 5ab + 6b^2$

6. $c^2 - cd - 30d^2$

7. $x^2 + 5xy - 24y^2$

8. $a^2 - 4ab + 4b^2$

9. $c^2 - 11cd + 30d^2$

10. $x^2 - 6xy + 8y^2$

11. $g^2 - gh - 42h^2$

12. $a^2 - ab - 20b^2$

13. $x^2 + 12xy + 32y^2$

14. $c^2 + 3cd - 40d^2$

15. $x^2 + 6xy - 27y^2$

16. $a^2 - 2ab - 48b^2$

17. $c^2 - 3cd - 28d^2$

18. $x^2 + xy - 6y^2$

5.8   Factoring the Difference of Two Squares

Let's give an example of a **perfect square**.

25 is a perfect square because $5 \times 5 = 25$
49 is a perfect square because $7 \times 7 = 49$

Any variable with an even exponent is a perfect square.

y^2 is a perfect square because $y \times y = y^2$
y^4 is a perfect square because $y^2 \times y^2 = y^4$

When two terms that are both perfect squares are subtracted, factoring those terms is very easy. To factor the difference of perfect squares, you use the square root of each term, a plus sign in the first factor, and a minus sign in the second factor.

Example 12:   Factor $4x^2 - 9$

This example has two terms which are both perfect squares, and the terms are subtracted.

Step 1: $(2x \quad 3)(2x \quad 3)$

Find the square root of each term.
Use the square roots in each of the factors.

Step 2: $(2x + 3)(2x - 3)$

Use a plus sign in one factor and a minus sign in the other factor.

Check: Multiply to check. $(2x + 3)(2x - 3) = 4x^2 - 6x + 6x - 9 = 4x^2 - 9$

The inner and outer terms add to zero.

Example 13:   Factor $81y^4 - 1$

Step 1: $(9y^2 + 1)(9y^2 - 1)$

Factor like the example above.
Notice, the second factor is also the difference of two perfect squares.

Step 2: $(9y^2 + 1)(3y + 1)(3y - 1)$

Factor the second term further.
Note: You cannot factor the sum of two perfect squares.

Check: Multiply in reverse to check your answer.
$(9y^2 + 1)(3y + 1)(3y - 1) = (9y^2 + 1)(9y^2 - 3y + 3y - 1) =$
$(9y^2 + 1)(9y^2 - 1) = 81y^4 + 9y^2 - 9y^2 - 1 = 81y^4 - 1$

Factor the following differences of perfect squares.

1. $64x^2 - 49$

2. $4y^4 - 25$

3. $9a^4 - 4$

4. $25c^4 - 9$

5. $64y^2 - 9$

6. $x^4 - 16$

7. $49x^2 - 4$

8. $4d^2 - 25$

9. $9a^2 - 16$

10. $100y^4 - 49$

11. $c^4 - 36$

12. $36x^2 - 25$

13. $25x^2 - 4$

14. $9x^4 - 64$

15. $49x^2 - 100$

16. $16x^2 - 81$

17. $9y^4 - 1$

18. $49c^2 - 25$

19. $25d^2 - 64$

20. $36a^4 - 49$

21. $16x^4 - 16$

22. $b^2 - 25$

23. $c^4 - 144$

24. $9y^2 - 4$

25. $81x^4 - 16$

26. $4b^2 - 36$

27. $9w^2 - 9$

28. $64a^2 - 25$

29. $49y^2 - 121$

30. $x^6 - 9$

5.9   Factoring Using Special Products

In Algebra, we use special products to factor expressions. There are only 6 formulas for special products and they are:

1. $x^2 + 2xy + y^2 = (x + y)^2$ Square of a sum
2. $x^2 - 2xy + y^2 = (x - y)^2$ Square of a difference
3. $x^2 - y^2 = (x + y)(x - y)$ Product of a sum and difference
4. $x^2 + (a + b)x + ab = (x + a)(x + b)$ Product of a sum/difference with two different numbers
5. $x^3 + 3x^2y + 3xy^2 + y^3 = (x + y)^3$ Cube of a sum
6. $x^3 - 3x^2y + 3xy^2 - y^3 = (x - y)^3$ Cube of a difference

Using the formulas for special products, we can quickly factor these problems.

Example 14: $x^2 + 6x + 9 = (x + 3)^2$

Example 15: $x^2 - 6x + 9 = (x - 3)^2$

Example 16: $x^2 - 9 = (x + 3)(x - 3)$

Example 17: $x^2 + 5x + 5 = (x + 3)(x + 2)$

Example 18: $x^3 + 9x^2 + 27x + 27 = (x + 3)^3$

Example 19: $x^3 - 9x^2 + 27x - 27 = (x - 3)^3$

Use formulas for special products to expand the binomials.

1. $(x + 2)^2$ 4. $(x - 9)^2$ 7. $(x + 4)(x + 5)$ 10. $(x + 4)^3$

2. $(x + 7)^2$ 5. $(x + 7)(x - 7)$ 8. $(x + 7)(x + 2)$ 11. $(x - 8)^3$

3. $(x - 6)^2$ 6. $(x + 5)(x - 5)$ 9. $(x + 6)^3$ 12. $(x - 1)^3$

Use what you know about special products to factor the expressions.

13. $x^2 + 6x + 9$ 19. $x^2 + 15x + 50$

14. $x^2 + 20x + 100$ 20. $x^2 + 11x + 28$

15. $x^2 - 14x + 49$ 21. $x^3 + 6x^2 + 12x + 8$

16. $x^2 - 18x + 81$ 22. $x^3 + 12x^2 + 48x + 64$

17. $x^2 - 1$ 23. $x^3 - 9x^2 + 27x - 27$

18. $x^2 - 121$ 24. $x^3 - 15x^2 + 75x - 125$

5.10   Simplifying Rational Expressions

We will use what we learned so far in this chapter to factor the terms in the numerator and the denominator when possible, then simplify the rational expression.

Example 20:   Simplify $\dfrac{c^2 - 25}{c^2 + 5c}$

Step 1:   The numerator is the difference of two perfect squares, so it can be easily factored as in the previous section. Use the square root of each of the terms in the parentheses, with a plus sign in one and a minus sign in the other.
$c^2 - 25 = (c - 5)(c + 5)$

Step 2:   Find the greatest common factor in the denominator and factor it out. In this case, it is the variable c.
$c^2 + 5c = c(c + 5)$

Step 3:   Simplify $\dfrac{c^2 - 25}{c^2 + 5c} = \dfrac{(c - 5)(c + 5)}{c(c + 5)} = \dfrac{c - 5}{c}$

Simplify the rational expressions. Check for perfect squares and common factors.

1. $\dfrac{25x^2 - 4}{5x^2 - 2x}$

2. $\dfrac{64c^2 - 25}{8c^2 + 5c}$

3. $\dfrac{36a^2 - 49}{6a^2 - 7a}$

4. $\dfrac{x^2 - 9}{x^2 + 3x}$

5. $\dfrac{9a^2 - 16}{3a^2 - 4a}$

6. $\dfrac{16x^2 - 81}{4x^2 - 9x}$

7. $\dfrac{49x^2 - 100}{7x^2 + 10x}$

8. $\dfrac{x^4 - 16}{x^2 + 2x}$

9. $\dfrac{4y^2 - 36}{2y^2 + 6y}$

10. $\dfrac{81y^4 - 16}{9y^2 + 4}$

11. $\dfrac{25x^4 - 225}{5x^2 + 15}$

12. $\dfrac{3y^3 + 9}{y^9 - 9}$

5.11   Adding Rational Expressions

Rational expressions are fractions that can have variables in the numerator and/or the denominator. Adding rational expressions is similar to adding fractions.

Example 21:   Add: $\dfrac{x-3}{x-2} + \dfrac{5}{x+3}$

Step 1:   Just like adding fractions, we must find a common denominator by the finding the least common multiple of the denominators. The least common multiple of $x-2$ and $x+3$ is $(x-2)(x+3)$.

Step 2:   Set up the algebra problem like a fraction problem and find the numerators.

$$\dfrac{x-3}{x-2} = \dfrac{(x-3)(x+3)}{(x-2)(x+3)}$$

$$+ \quad \dfrac{5}{x+3} = \dfrac{5x-10}{(x+3)(x-2)}$$

Step 3:   Add: $\dfrac{(x-3)(x+3)+5x-10}{(x+3)(x-2)}$

Step 4:   Simplify: $\dfrac{x^2-9+5x-10}{(x+3)(x-2)} = \dfrac{x^2+5x-19}{(x+3)(x-2)}$

Add the following rational expressions.

1. $\dfrac{x^2-9x-22}{x^2-8x-33} + \dfrac{x^2+12x+35}{x^2+10x+21}$

2. $\dfrac{1}{x^3} + \dfrac{x-9}{x^2}$

3. $\dfrac{1}{2x} + \dfrac{x^2-3}{2x}$

4. $\dfrac{y^3}{x} + \dfrac{x^3}{y}$

5. $\dfrac{(x-2)(x+13)}{x^2-4} + \dfrac{(x-3)(x+1)}{x^2-x-6}$

6. $\dfrac{a}{b^2} + \dfrac{b^3}{a-b}$

7. $\dfrac{5}{y^3} + \dfrac{7}{y^3}$

8. $\dfrac{12b}{a^2} + \dfrac{31c}{a^2}$

9. $\dfrac{y}{x^2} + \dfrac{3}{y^2} + \dfrac{x}{x^2y^2}$

10. $\dfrac{r}{q^{12}} + \dfrac{s}{r^{15}} + \dfrac{q}{s^5}$

11. $\dfrac{x}{x-y} + \dfrac{y}{x-y}$

12. $\dfrac{9}{ba} + \dfrac{2}{b}$

5.12   Subtracting Rational Expressions

Subtracting rational expressions is similar to subtracting fractions.

Example 22:   Subtract: $\dfrac{x^2}{5} - \dfrac{9}{x+3}$

Step 1:   Just like adding we must find a common denominator. The common denominator of the rational expressions is $5\left(x+3\right)$.

Step 2:   Set up the algebra problem like a fraction problem and find the numerators.

$$\frac{x^2}{5} = \frac{x^2\left(x+3\right)}{5\left(x+3\right)}$$

$$-\ \frac{9}{x+3} = \frac{45}{5\left(x+3\right)}$$

Step 3:   Subtract: $\dfrac{x^2\left(x+3\right)-45}{5\left(x+3\right)}$

Step 4:   Simplify: $\dfrac{x^2\left(x+3\right)-45}{5\left(x+3\right)} = \dfrac{x^3+3x^2-45}{5\left(x+3\right)}$

Subtract the following rational expressions.

1. $\dfrac{x}{3y} - \dfrac{2}{3y}$

2. $\dfrac{7}{5c} - \dfrac{2}{c}$

3. $\dfrac{x}{y^2} - \dfrac{x^2}{y^2}$

4. $\dfrac{x+1}{y+1} - \dfrac{y+1}{x+1}$

5. $\dfrac{1}{x} - \dfrac{2x^2}{y}$

6. $\dfrac{2a^2}{bc} - \dfrac{a^5}{bc}$

7. $\dfrac{d-e}{a+b} - \dfrac{5}{a+b}$

8. $\dfrac{x}{5y^2} - \dfrac{y}{5x^2}$

9. $\dfrac{c^3}{3b} - \dfrac{b^2}{4a}$

10. $\dfrac{2x+5}{y} - \dfrac{5x^3-x^2}{y}$

11. $\dfrac{2d}{3c^5} - \dfrac{a^5}{2b^2}$

12. $\dfrac{5x}{y} - \dfrac{2x}{y}$

5.13   Multiplying Rational Expressions

Multiplying rational expressions is similar to multiplying fractions because a rational expression is a fraction.

Example 23:   Multiply: $\dfrac{x^4 - x^3}{y^3} \times \dfrac{y^4}{x^2}$

Step 1:   When you multiply a rational expression, you must multiply the numerators together and multiply the denominators together.

$$\frac{x^4 - x^3}{y^3} \times \frac{y^4}{x^2} = \frac{(x^4 - x^3) \times y^4}{y^3 \times x^2} = \frac{x^4 y^4 - x^3 y^4}{x^2 y^3}$$

Step 2:   Simplify the resulting rational expression. You can factor $x^3 y^4$ out of the numerator.

$$\frac{x^4 y^4 - x^3 y^4}{x^2 y^3} = \frac{x^3 y^4 (x - 1)}{x^2 y^3}$$

Step 3:   You can also cancel $x^2 y^3$ because it is in the numerator and denominator of the expression.

$$\frac{x^3 y^4 (x - 1)}{x^2 y^3} = \frac{xy (x - 1)}{1} = xy (x - 1)$$

Therefore, $\dfrac{x^4 - x^3}{y^3} \times \dfrac{y^4}{x^2} = xy (x - 1)$.

Multiply the following rational expressions.

1. $\dfrac{-b}{2a} \times \dfrac{-a}{3b}$

2. $\dfrac{4x}{y} \times \dfrac{1}{2y}$

3. $\dfrac{5a}{3b} \times \dfrac{4c}{3b}$

4. $\dfrac{a + b}{a - b} \times \dfrac{c + b}{c - b}$

5. $\dfrac{c}{a^2} \times \dfrac{c - a}{b - c}$

6. $\dfrac{y^5}{x^3} \times \dfrac{y^2 + 2y + 1}{x^2 - y}$

7. $\dfrac{x^2 - 2x - 3}{x^2 - 5x - 14} \times \dfrac{x^2 - 2x - 35}{x^2 + 6x - 27}$

8. $\dfrac{9}{x} \times \dfrac{x^5}{y}$

9. $\dfrac{a^2 - b}{a - b^2} \times \dfrac{7c}{b}$

10. $\dfrac{5}{x^2 y} \times \dfrac{4}{xy^2}$

11. $\dfrac{5x^3}{2} \times \dfrac{2x^3}{5}$

12. $\dfrac{x + 7}{x - 1} \times \dfrac{x - 3}{x + 5}$

13. $\dfrac{3c}{a^2} \times \dfrac{2ba}{c}$

14. $\dfrac{a^5 - a^2}{b^3} \times \dfrac{c^4}{a^2}$

15. $\dfrac{b - 17}{c^2 + 2} \times \dfrac{a^3}{b^3} \times \dfrac{c^3}{b - 17}$

16. $\dfrac{hk}{47} \times \dfrac{16}{m}$

5.14   Dividing Rational Expressions

Dividing rational expressions is similar to dividing fractions because a rational expression is a fraction.

Example 24:   Divide: $\dfrac{2a^3}{b} \div \dfrac{5}{b^2}$

Step 1:   Just like dividing fractions, you must flip the second expression to get its reciprocal, then multiply.

$$\frac{2a^3}{b} \div \frac{5}{b^2} = \frac{2a^3}{b} \times \frac{b^2}{5}$$

Step 2:   Multiply the two rational expressions together just like you did in the previous section.

$$\frac{2a^3}{b} \times \frac{b^2}{5} = \frac{2a^3 \times b^2}{b \times 5} = \frac{2a^3b^2}{5b}$$

Step 3:   Simplify: $\dfrac{2a^3b^2}{5b} = \dfrac{2a^3b}{5}$

Therefore, $\dfrac{2a^3}{b} \div \dfrac{5}{b^2} = \dfrac{2a^3b}{5}$.

Divide the following rational expressions.

1. $\dfrac{x}{y} \div \dfrac{y}{x}$

2. $\dfrac{2x}{y^3} \div \dfrac{x}{y}$

3. $\dfrac{x-1}{y+2} \div \dfrac{y-5}{x+7}$

4. $\dfrac{a-2}{c-1} \div \dfrac{a-2}{c+3}$

5. $\dfrac{x^2+2x+1}{y^2+8y+15} \div \dfrac{x^2+10x+9}{y^2+10y+21}$

6. $\dfrac{a}{c} \div \dfrac{c}{b}$

7. $\dfrac{c}{a} \div \dfrac{c}{b}$

8. $\dfrac{2x+3}{x+3} \div \dfrac{x+3}{2x+3} \div \dfrac{x}{y}$

9. $\dfrac{ac}{b^2} \div \dfrac{c^2}{ab}$

10. $\dfrac{b^2}{a} \div \dfrac{a^2}{c}$

11. $\dfrac{c^2}{3b} \div \dfrac{16}{a}$

12. $\dfrac{20cd}{b} \div \dfrac{2ac}{d}$

13. $\dfrac{x+4}{x+2} \div \dfrac{x-1}{x-3}$

14. $\dfrac{y}{3x^2} \div \dfrac{2y}{x^3}$

15. $\dfrac{y^3-1}{x^2+1} \div \dfrac{x-7}{y+2}$

16. $\dfrac{200a}{7b} \div \dfrac{10a}{b}$

Chapter 5 Review

Factor the following polynomials completely.

1. $8x - 18$

2. $6x^2 - 18x$

3. $16b^3 + 8b$

4. $15a^3 + 40$

5. $20y^6 - 12y^4$

6. $5a - 15a^2$

7. $4y^2 - 36$

8. $25a^4 - 49b^2$

9. $3ax + 3ay + 4x + 4y$

10. $ax - 2x + ay - 2y$

11. $2bx + 2x - 2by - 2y$

12. $2b^2 - 2b - 12$

13. $7x^3 + 14x - 3x^2 - 6$

14. $3a^3 + 4a^2 + 9a + 12$

15. $27y^2 + 42y - 5$

16. $12b^2 + 25b - 7$

17. $c^2 + cd - 20d^2$

18. $x^2 - 4xy - 21y^2$

19. $6y^2 + 30y + 36$

20. $2b^2 + 6b - 20$

21. $16b^4 - 81d^4$

22. $9w^2 - 54w - 63$

23. $m^2p^2 - 5mp + 2m^2p - 10m$

24. $12x^2 + 27x$

25. $2xy - 36 + 8y - 9x$

26. $2a^4 - 32$

27. $21c^2 + 41c + 10$

28. $x^2 - y + xy - x$

29. $2b^3 - 24 + 16b - 3b^2$

30. $5 - 2a - 25a^2 + 10a^3$

Simplify the following rational expressions by performing the appropriate operation.

31. $\dfrac{b}{a} + \dfrac{b}{a}$

32. $\dfrac{x}{y} + \dfrac{y}{x}$

33. $\dfrac{x}{x+y} - \dfrac{y}{x+y}$

34. $\dfrac{3x+5}{x+1} - \dfrac{3x-1}{x-1}$

35. $\dfrac{-b}{ac} \times \dfrac{3b}{a-2} \times \dfrac{c}{b+1}$

36. $\dfrac{x}{y^3} \times \dfrac{3y^2}{x^2}$

37. $\dfrac{-b}{5ac} \div \dfrac{d}{b^2}$

38. $\dfrac{5}{a+3} \div \dfrac{7a+21}{b^2+b}$

Use your knowledge of special products to factor the following expressions.

39. $x^2 + 12x + 36$

40. $x^2 - 8x + 16$

41. $x^2 - 4$

42. $x^2 + 8x + 12$

43. $x^3 + 6x^2 + 12x + 8$

44. $x^3 - 12x^2 + 48x - 64$

45. $x^3 + 18x^2 + 108x + 216$

46. $x^3 + 12x^2 + 48x + 64$

47. $x^3 + 21x^2 + 147x + 343$

Chapter 5 Test

1. What is the greatest common factor of $4x^3$ and $8x^2$?

   (A) $4x^2$
   (B) $4x$
   (C) x^2
   (D) $8x$

2. Factor: $8x^4 - 7x^2 + 4x$

   (A) $4x\left(2x^3 - 7x + 4\right)$
   (B) $x\left(8x^4 - 7x^2 + 4x\right)$
   (C) $x\left(8x^3 - 7x + 4\right)$
   (D) $4x\left(2x^3 - 7x + 1\right)$

3. Factor by grouping:

   $xy + 2x + 3y + 6$

   (A) $(x + 2)(y + 3)$
   (B) $(x + 3)(y + 2)$
   (C) $x(y + 2) + 3(y + 2)$
   (D) This problem cannot be factored.

4. Factor: $x^2 + 6x + 8$

   (A) $(x + 2)(x + 4)$
   (B) $(x + 1)(x + 8)$
   (C) $(x - 2)(x - 4)$
   (D) $(x - 1)(x - 8)$

5. Divide: $\dfrac{5}{2x} \div \dfrac{5x}{y}$

   (A) $\dfrac{y}{2x}$

   (B) $\dfrac{25}{2x^2y}$

   (C) $\dfrac{y}{2x^2}$

   (D) $\dfrac{25x}{2y}$

6. Simplify the following rational expression:

   $$\dfrac{36x^4 - 16}{6x^3 + 4x}$$

   (A) $\dfrac{\left(6x^2 - 4\right)\left(6x^2 + 4\right)}{x}$

   (B) $\dfrac{6x^2 - 4}{x}$

   (C) $6x^2 - 4$

   (D) $\dfrac{(3x - 2)(3x + 2)}{x}$

7. Simplify the following rational expression:

   $$\dfrac{c^2 + 10c + 24}{c^3 + 4c^2}$$

   (A) $\dfrac{c + 6}{c}$

   (B) $\dfrac{c + 4}{c^2}$

   (C) $\dfrac{c + 6}{c^2}$

   (D) $\dfrac{(c + 6)(c + 4)}{c^2(c + 4)}$

8. Add: $\dfrac{3x^3}{y} + \dfrac{3y^3}{x}$

   (A) $\dfrac{9x^3y^3}{xy}$

   (B) $\dfrac{3x^4 + 3y^4}{xy}$

   (C) $\dfrac{3x^3 + 3y^3}{x + y}$

   (D) $\dfrac{3x^4 + 3y^4}{x + y}$

9. Add: $\dfrac{11x-y}{y}+\dfrac{2}{xy^2}$

(A) $\dfrac{11x^2y-xy^2+2}{xy^3}$

(B) $\dfrac{11x^2y-xy^2-2}{xy^2}$

(C) $\dfrac{22x-2y}{y+xy^2}$

(D) $\dfrac{11x^2y-xy^2+2}{xy^2}$

10. Subtract: $\dfrac{x+2}{x^2-4}-\dfrac{x}{x-2}$

(A) $\dfrac{1-x}{x-2}$

(B) $\dfrac{x}{x^2-4}$

(C) $\dfrac{2x-2}{x^2-4}$

(D) $\dfrac{x+2}{x^2-x-6}$

11. Multiply: $\dfrac{x+3}{2x}\times\dfrac{x(y-1)}{y}$

(A) $\dfrac{xy-x+3y-3}{2x}$

(B) $\dfrac{xy-x+3y-3}{2xy}$

(C) $\dfrac{xy-x+3y-3}{xy}$

(D) $\dfrac{xy-x+3y-3}{2y}$

12. Factor: $2x^2-2x-84$

(A) $(2x+7)(x-12)$
(B) $(2x-12)(x+7)$
(C) $(2x-7)(x+12)$
(D) $(2x+12)(x-7)$

13. Subtract: $\dfrac{1}{7x}-\dfrac{13y}{x}$

(A) $\dfrac{14y}{8x}$

(B) $-\dfrac{13y}{7x}$

(C) $\dfrac{1-13y}{6x}$

(D) $\dfrac{1-91y}{7x}$

14. Multiply: $\dfrac{12y}{x}\times\dfrac{y^3}{x}$

(A) $\dfrac{12y^4}{x^2}$

(B) $\dfrac{12y}{x^3}$

(C) $\dfrac{36y^4}{x^2}$

(D) $\dfrac{12y^4}{x}$

15. Divide: $\dfrac{x-2}{y}\div\dfrac{7}{x+3}$

(A) $\dfrac{7x-14}{y+x+3}$

(B) $\dfrac{x^2+x-6}{7y}$

(C) $\dfrac{7x-14}{xy+3y}$

(D) $\dfrac{x^2+x-6}{y}$

16. Factor: $4x^2-64$

(A) $(x-8)(x+8)$
(B) $(4x-8)(4x+8)$
(C) $(2x-16)(2x+16)$
(D) $(2x-8)(2x+8)$

17. Factor: $x^2 + 22x + 121$

  (A) $(x + 12)^2$
  (B) $(x - 11)^2$
  (C) $(x + 7)^2$
  (D) $(x + 11)^2$

18. Factor: $x^2 - 24x + 144$

  (A) $(x + 12)^2$
  (B) $(x - 12)^2$
  (C) $(x + 12)(x - 12)$
  (D) $(x + 12)^3$

19. Factor: $x^2 - 36$

  (A) $(x + 7)(x + 4)$
  (B) $(x + 4)(x - 9)$
  (C) $(x + 6)(x - 6)$
  (D) $(x + 6)(x + 6)$

20. Factor: $x^3 + 18x^2 + 108x + 216$

  (A) $(x + 6)^3$
  (B) $(x + 4)^3$
  (C) $(x + 8)^3$
  (D) $(x - 6)^3$

21. Factor: $x^3 - 27x^2 + 243x - 729$

  (A) $(x - 11)^3$
  (B) $(x - 9)^3$
  (C) $(x - 7)^3$
  (D) $(x - 8)^3$

22. What are the factors of $x^2 + 10x + 25$?

  (A) $(x + 5)^2$
  (B) $(x - 5)^2$
  (C) $(x - 5)(x + 5)$
  (D) $(x + 5)(x + 2)$

23. What are the factors of $x^2 + 11x + 30$?

  (A) $(x + 6)(x + 5)$
  (B) $(x + 10)(x + 3)$
  (C) $(x + 6)^2$
  (D) $(x + 15)(x + 2)$

24. What are the factors of $x^3 - 3x^2 + 3x - 1$?

  (A) $(x + 1)^3$
  (B) $(x + 3)^3$
  (C) $(x - 3)^3$
  (D) $(x - 1)^3$

25. What are the factors of $x^3 + 21x^2 + 147x + 343$?

  (A) $(x + 7)^3$
  (B) $(x - 7)^3$
  (C) $(x - 3)^3$
  (D) $(x + 9)^3$

Chapter 6
Solving Quadratic Equations

This chapter covers the following Georgia Performance Standards:

MM1A	Algebra	MM1A3a

6.1   Solving Quadratic Equations

In the previous chapter, we factored polynomials such as $y^2 - 4y - 5$ into two factors:

$$y^2 - 4y - 5 = (y + 1)(y - 5)$$

In this chapter, we learn that any equation that can be put in the form $ax^2 + bx + c = 0$ is a quadratic equation if a, b, and c are real numbers and $a \neq 0$. $ax^2 + bx + c = 0$ is the standard form of a quadratic equation. To solve these equations, follow the steps below.

Example 1:   Solve $y^2 - 4y - 5 = 0$

Step 1:   Factor the left side of the equation.

$$y^2 - 4y - 5 = 0$$
$$(y + 1)(y - 5) = 0$$

Step 2:   If the product of these two factors equals zero, then the two factors individually must be equal to zero. Therefore, to solve, we set each factor equal to zero.

$$
\begin{array}{cc}
(y + 1) = 0 & (y - 5) = 0 \\
\underline{-1 \quad -1} & \underline{+5 \quad +5} \\
y = -1 & y = 5
\end{array}
$$

The equation has two solutions: $y = -1$ and $y = 5$.

Check: To check, substitute each solution into the original equation.

When $y = -1$, the equation becomes: When $y = 5$, the equation becomes:
$$(-1)^2 - (4)(-1) - 5 = 0 \qquad\qquad 5^2 - (4)(5) - 5 = 0$$
$$1 + 4 - 5 = 0 \qquad\qquad 25 - 20 - 5 = 0$$
$$0 = 0 \qquad\qquad 0 = 0$$

Both solutions produce true statements.
The solution set for the equation is $\{-1, 5\}$.

Solve each of the following quadratic equations by factoring and setting each factor equal to zero. Check by substituting answers back in the original equation.

1. $x^2 + x - 6 = 0$

2. $y^2 - 2y - 8 = 0$

3. $a^2 + 2a - 15 = 0$

4. $y^2 - 5y + 4 = 0$

5. $b^2 - 9b + 14 = 0$

6. $x^2 - 3x - 4 = 0$

7. $y^2 + y - 20 = 0$

8. $d^2 + 6d + 8 = 0$

9. $y^2 - 7y + 12 = 0$

10. $x^2 - 3x - 28 = 0$

11. $a^2 - 5a + 6 = 0$

12. $b^2 + 3b - 10 = 0$

13. $a^2 + 7a - 8 = 0$

14. $x^2 + 3x + 2 = 0$

15. $x^2 - x - 42 = 0$

16. $a^2 + a - 6 = 0$

17. $b^2 + 7b + 12 = 0$

18. $y^2 + 2y - 15 = 0$

19. $a^2 - 3a - 10 = 0$

20. $x^2 + 10x + 16 = 0$

21. $x^2 - 4x - 12 = 0$

6.2   Solving Perfect Squares

When the square root of a constant, variable, or polynomial results in a constant, variable, or polynomial without irrational numbers, the expression is a **perfect square**. Some examples are 49, x^2, and $(x-2)^2$.

Example 2:   Solve the perfect square for x. $(x-5)^2 = 0$

Step 1:   Take the square root of both sides.
$$\sqrt{(x-5)^2} = \sqrt{0}$$
$$(x-5) = 0$$

Step 2:   Solve the equation.
$$(x-5) = 0$$
$$x - 5 + 5 = 0 + 5$$
$$x = 5$$

Example 3:   Solve the perfect square for x. $(x-5)^2 = 64$

Step 1:   Take the square root of both sides.
$$\sqrt{(x-5)^2} = \sqrt{64}$$
$$(x-5) = \pm 8$$
$$(x-5) = 8 \text{ and } (x-5) = -8$$

Step 2:   Solve the two equations.
$$(x-5) = 8 \qquad \text{and} \quad (x-5) = -8$$
$$x - 5 + 5 = 8 + 5 \quad \text{and} \quad x - 5 + 5 = -8 + 5$$
$$x = 13 \qquad\quad \text{and} \quad x = -3$$

Solve the perfect square for x.

1. $(x-2)^2 = 0$

2. $(x+1)^2 = 0$

3. $(x+11)^2 = 0$

4. $(x-4)^2 = 0$

5. $(x-1)^2 = 0$

6. $(x+8)^2 = 0$

7. $(x+3)^2 = 4$

8. $(x-5)^2 = 16$

9. $(x-10)^2 = 100$

10. $(x+9)^2 = 9$

11. $(x-4.5)^2 = 25$

12. $(x+7)^2 = 36$

13. $(x+2)^2 = 49$

14. $(x-1)^2 = 4$

15. $(x+8.9)^2 = 49$

16. $(x-6)^2 = 81$

17. $(x-12)^2 = 121$

18. $(x+2.5)^2 = 64$

Chapter 6 Review

Factor and solve each of the following quadratic equations.

1. $a^2 - a - 30 = 0$

2. $x^2 - x = 6$

3. $y^2 = 21 - 4y$

4. $y^2 - 7y + 8 = 16$

5. $b^2 + 2b - 8 = 0$

6. $y^2 + 3y - 18 = 0$

7. $x^2 + 10x - 11 = 0$

8. $y^2 - 14y + 40 = 0$

9. $b^2 + 9b + 18 = 0$

10. $y^2 - 12y - 13 = 0$

11. $a^2 - 8a - 48 = 0$

12. $x^2 + 2x - 63 = 0$

Chapter 6 Test

1. Solve for x. $x^2 - 3x - 18 = 0$

   (A) $\{-6, 3\}$
   (B) $\{6, -3\}$
   (C) $\{-9, 2\}$
   (D) $\{9, -2\}$

2. What are the values of x in the quadratic equation? $x^2 + 2x - 15 = x - 3$

   (A) $\{-4, 3\}$
   (B) $\{-3, 4\}$
   (C) $\{-3, 5\}$
   (D) Cannot be determined

3. Solve the equation $(x + 9)^2 = 49$

   (A) $x = -9, 9$
   (B) $x = -9, 7$
   (C) $x = -16, -2$
   (D) $x = -7, 7$

4. Solve the equation $c^2 + 8c - 9 = 0$.

   (A) $c = \{1, -9\}$
   (B) $c = \{-1, 9\}$
   (C) $c = \{3, 3\}$
   (D) $c = \{-3, -3\}$

5. Solve the following equation:

   $x^2 = 3x$

   (A) $x = \{0, 1\}$
   (B) $x = \{3, 1\}$
   (C) $x = \{0, 3\}$
   (D) $x = \{3, -3\}$

Chapter 7
Triangles

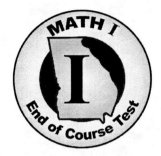

This chapter covers the following Georgia Performance Standards:

MM1G	Geometry	MM1G3a
		MM1G3b
		MM1G3c
		MM1G3e

7.1 Types of Triangles

A triangle is an **equilateral triangle** if all of its sides and angles are equal. A triangle is an **isosceles triangle** if two of its sides and the angles opposite those sides are equal. A triangle is a **right triangle** if one of its angles equals 90°.

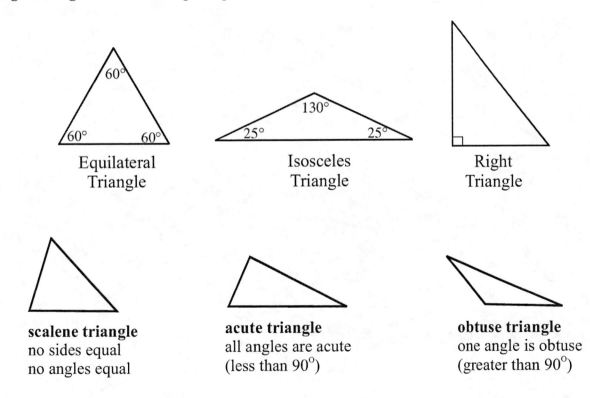

Equilateral Triangle

Isosceles Triangle

Right Triangle

scalene triangle
no sides equal
no angles equal

acute triangle
all angles are acute
(less than 90°)

obtuse triangle
one angle is obtuse
(greater than 90°)

7.2   Interior Angles of a Triangle

The three interior angles of a triangle always add up to 180°.

Example 1:

$$45° + 45° + 90° = 180° \qquad 30° + 60° + 90° = 180° \qquad 60° + 60° + 60° = 180°$$

Example 2:   Find the missing angle in the triangle.

Solution:

$$
\begin{aligned}
20° + 125° + x &= 180° \\
-20° \; -125° \qquad &\quad -20° \; -125° \\
x &= 180° - 20° - 125° \\
x &= 35°
\end{aligned}
$$

Subtract 20° and 125° from both sides to get x by itself.

The missing angle is 35°.

Find the missing angle in the triangles.

1.

2.

3.

4.

5.

6.

7.

8.

9.

Find the missing angles in the triangles.

10.

11.

12.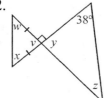

7.3   Exterior Angles

The **exterior angle** of a triangle is always equal to the sum of the opposite interior angles.

Example 3:   Find the measure of $\angle x$ and $\angle y$.

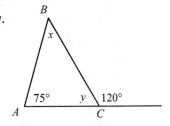

Step 1:   Using the rule for exterior angles,
$120° = \angle A + \angle B$
$120° = 75° + x$
$45° = x$

Step 2:   The sum of the interior angles of a triangle equals 180°, so
$180° = 75° + 45° + y$
$60° = y$

Find the measures of x and y.

1.

3.

5.

2.

4.

6.
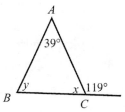

Find the measures of the angles.

7.

8.

9.

7.4   Side and Angle Relationships

The side-angle inequality theorem states that the longest side must be opposite the largest angle in the triangle. Also, the smallest angle must be opposite the shortest side in the triangle.

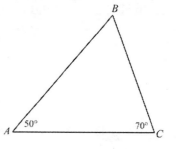

Example 4:   Given $\triangle ABC$, list the sides from shortest to longest.

Step 1:   First, we need to find the measure of $\angle B$.
The sum of the measures of the interior angles of a triangle is 180°, so
$m\angle A + m\angle B + m\angle C = 180°$.
$50° + m\angle B + 70° = 180°$
$m\angle B = 60°$.

Step 2:   By theorem, we know that if one angle of a triangle is larger than another angle, then the side **opposite** the larger angle is longer than the side opposite the smaller angle. Similarly, the side opposite the smallest angle of a triangle is the shortest side.

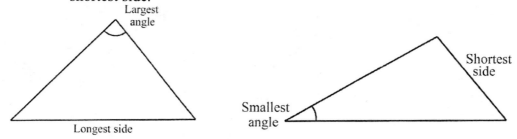

Therefore, in $\triangle ABC$, the shortest side is $\overline{BC}$, which is across from the 50° angle. The longest side is $\overline{AB}$, which is across from the 70° angle. In order from shortest to longest, then, the sides are $\overline{BC}$, $\overline{AC}$, and $\overline{AB}$. $\overline{BC} < \overline{AC} < \overline{AB}$

Solve the following problems.

1. List the sides of the triangle in order from longest to shortest.

2. List the angles of the triangle in order from smallest to largest.

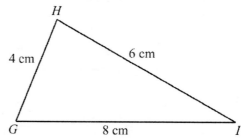

7.5   Triangle Inequality Theorem

The triangle inequality theorem states that the sum of the measure of any two sides in a triangle must be greater than the measure of the third side.

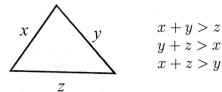

$$x + y > z$$
$$y + z > x$$
$$x + z > y$$

Example 5:   Determine whether or not it is possible to create a triangle with sides of 1 units, 5 units, and 7 units.

Step 1:   First, you must set up three inequalities. Remember the sum of any two sides of a triangle must be greater than the third side.

$$1 + 5 > 7 \qquad 1 + 7 > 5 \qquad 5 + 7 > 1$$

Step 2:   Determine if the inequalities are true.

$1 + 5 > 7$	$1 + 7 > 5$	$5 + 7 > 1$
$6 > 7$	$8 > 5$	$12 > 1$
False	True	True

The number 6 is not greater than 7, so a triangle cannot be formed using the sides given.
(All three inequalities must be true in order to create a triangle.)

Determine whether or not it is possible to create a triangle given the following measures of sides. Write yes if it is possible to form a triangle with the given measures of sides or write no if it is not possible.

1. 7, 8, 13

2. 2, 5, 9

3. 10, 8, 15

4. 6, 9, 20

5. 101, 89, 150

6. 1, 2, 4

7. 7, 7, 14

8. 21, 15, 29

9. 11, 9, 17

The triangle inequality theorem can also be used to find all possible lengths of a side of a triangle.

Example 6:   The length of two sides of a traingle are $3n$ and $n+1$ units, where $n > 0$. Find all the possible lengths, x, for the third side of the triangle.

Step 1:   We must find the range (maximum and minimum) of values x can be.

We know that the two sides in the problem must add up to be greater than the third side, x.

First, we will find the greatest amount that x can be (the maximum).

$3n + n + 1 > x$

$4n + 1 > x$

So, $x < 4n + 1$.

Step 2:   Now, we will find the minimum x can be. To do this, we must know the largest side out of the two sides given.

Since $n > 0$, we know that $3n$ is larger than $n + 1$.

Now, we will add x to the smaller side. This sum must be greater than the larger side.

$x + n + 1 > 3n$

Now solve for x.

$x + n + 1 - n - 1 > 3n - n - 1$

$x > 2n - 1$

Since $x > 2n - 1$ and $x < 4n + 1$, all the possible lengths for the third side of the triangle are $2n - 1 < x < 4n + 1$.

Find all the possible lengths, x, for the third side of the triangle.

1. The length of two sides of a traingle are n and $n - 2$ units, where $n > 2$.

2. The length of two sides of a traingle are $2n$ and $3n + 1$ units, where $n > 0$.

3. The length of two sides of a traingle are n and $n + 1$ units, where $n > 0$.

4. The length of two sides of a traingle are $4n$ and $n - 3$ units, where $n > 3$.

5. The length of two sides of a traingle are $3n - 3$ and $5n$ units, where $n > 3$.

6. The length of two sides of a traingle are n and $2n + 2$ units, where $n > 0$.

7. The length of two sides of a traingle are $3n$ and $n - 3$ units, where $n > 3$.

8. The length of two sides of a traingle are n and $n - 5$ units, where $n > 5$.

9. The length of two sides of a traingle are $n - 4$ and $8n$ units, where $n > 4$.

10. The length of two sides of a traingle are $4n$ and $2n + 1$ units, where $n > 0$.

7.6   Exterior-Angle Inequality

The exterior angle inequality states that an exterior angle of a triangle is equal in measure to the sum of the two non-adjacent interior angles of the triangle.

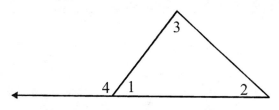

$$m\angle 4 = m\angle 2 + m\angle 3$$

Example 7:   Find the measure of x.

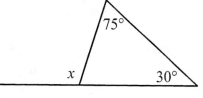

Step 1:   Since you know that x equals the sum of the measures of the two non-adjacent interior angles of the triangle, set up your equation.
$$x = 75° + 30°$$

Step 2:   Solve for x. $x = 75° + 30° = 105°$

Example 8:   Find the measure of x.

Step 1:   Set up your equation using the exterior angle inequality.
$$130° = 80° + x$$

Step 2:   Solve for x. $x = 130° - 80° = 50°$

Find the value of x.

1.

3.

5.

2.

4.

6.

7.7   Congruence Postulates

Some triangles are congruent and some are similar. There are specific principles that determine whether are not two triangles are congruent or similar. Two triangles are congruent, denoted by the symbol $\cong$, if all of the corresponding sides and angles of each triangle are equal to each other.

Side-Side-Side (SSS) Congruence Postulate	If three sides of one triangle are congruent to three sides of another triangle, then the two triangles are congruent.	If $\overline{AB} \cong \overline{xy}$, and $\overline{BC} \cong \overline{yz}$, and $\overline{AC} \cong \overline{xz}$, then $\triangle ABC \cong \triangle xyz$
Side-Angle-Side (SAS) Congruence Postulate	If two sides and the included angle of one triangle are congruent to two sides and the included angle of the second triangle, then the two triangles are congruent.	If $\overline{AB} \cong \overline{xy}$, and $\angle A \cong \angle x$, and $\overline{AC} \cong \overline{xz}$, then $\triangle ABC \cong \triangle xyz$
Angle-Side-Angle (ASA) Congruence Postulate	If two angles and the included side of one triangle are congruent to two angles and the included side of the second triangle, then the two triangles are congruent.	If $\angle B \cong \angle y$, and $\angle C \cong \angle z$, and $\overline{BC} \cong \overline{yz}$, then $\triangle ABC \cong \triangle xyz$
Angle-Angle-Side (AAS) Congruence Theorem	If two angles and a nonincluded side of one triangle are congruent to two angles and the corresponding side of a second triangle, then the two triangles are congruent.	If $\angle A \cong \angle x$, and $\angle C \cong \angle z$, and $\overline{BC} \cong \overline{yz}$, then $\triangle ABC \cong \triangle xyz$
Hypotenuse-Leg (HL) Theorem	If a hypotenuse and a leg of one right triangle are congruent to a hypotenuse and a leg of another right triangle, then the two triangles are congruent.	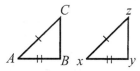 If $\overline{AC} \cong \overline{xz}$, and $\overline{AB} \cong \overline{xy}$, then $\triangle ABC \cong \triangle xyz$

7.8   Points of Concurrency

There are four special points in a triangle called the **centers** of a triangle and they are:

1. **Centroid** - point of concurrency of the medians

2. **Orthocenter** - point of concurrency of the altitudes

3. **Incenter** - point of concurrency of the angle bisectors

4. **Circumcenter** - point of concurrency of the perpendicular bisectors of the sides

Example 9:   The star denotes the **centroid** of $\triangle ABC$.

The centroid is the point where the three **medians** of the triangle intersect.

Example 10:   The star denotes the **orthocenter** of $\triangle ABC$.

The orthocenter is the point at which the three **altitudes** of the triangle intersect.

Example 11:   The star denotes the **incenter** (the center of a circle inscribed in a triangle) of $\triangle ABC$.

The incenter is located at the point of intersection of the triangle's **angle bisectors**.

Example 12:   The star denotes the **circumcenter** (the center of a circle circumscribed around the triangle).

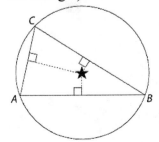

The circumcenter is the point of intersection of the triangle's **perpendicular bisectors**.

Example 13:   Below is a chart that lists the type of triangle and the concurrency location.

Acute   All four points are inside the triangle.

Obtuse   The incenter and centroid are inside while the orthocenter and circumcenter are outside.

Right   The incenter and centroid are inside. The orthocenter is on the 90° angle and the circumcenter is on the center of the hypotenuse.

Fill in the chart giving the location (in/out/on).

	Centroid	Orthocenter	Incenter	Circumcenter
Acute				
Right				
Obtuse				

True/False.

1. The altitudes meet at the circumcenter.

2. The medians meet at the incenter.

3. The medians meet at the centroid.

4. The perpendicular bisectors of the sides meet at the orthocenter.

Chapter 7 Review

1. Find the missing angle.

2. What is the largest angle of $\triangle ABC$?

3. Find x.

4. What type of triangle has only two equal sides?

5. Find the measures of x and y.

6. Find the measures of x and y.

For questions 7–10, determine which inequality theorem is shown: side-angle inequality, triangle inequality, or exterior-angle inequality.

7. $\overline{AB} > \overline{BC}$

8. $115° = 55° + x$

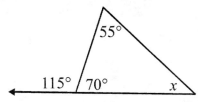

9. $m\angle A > m\angle C$

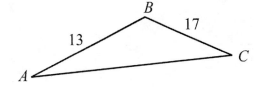

10. $\overline{AB} + \overline{BC} > \overline{AC}$

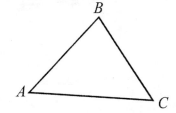

Determine whether the statements about points of concurrency below are true or false.

11. The circumcenter is also the center of the circle inscribed in a triangle.

12. All four points lie within a right triangle.

13. The incenter is where the angle bisectors meet.

14. The orthocenter is always inside the triangle.

15. In a right triangle, the circumcenter is on the middle of the hypotenuse.

16. The centroid is where the medians meet.

17. The centroid is always inside the triangle.

18. The incenter and circumcenter always lie within a triangle.

Determine if the following triangles are congruent. If they are congruent, state the theorem, SSS, SAS, ASA, AAS, or HL, used to prove they are congruent.

19.

22.

20.

23.

21.

24.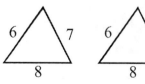

Chapter 7 Test

1. What is the measure of missing angle?

   (A) 225°
   (B) 45°
   (C) 75°
   (D) 30°

2. By what theorem can you prove the two triangles congruent?

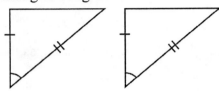

   (A) SAS
   (B) ASA
   (C) SSA
   (D) AAS

3. What is the measure of y?

   (A) 93°
   (B) 87°
   (C) 37°
   (D) Cannot be determined

4. Which two points are always inside the triangle?

   (A) Orthocenter & Centroid
   (B) Centroid & Incenter
   (C) Incenter & Orthocenter
   (D) Circumcenter & Centroid

5. What is the measure of the two missing angles?

   (A) $x = 26°, y = 133°$
   (B) $x = 47°, y = 112°$
   (C) $x = 112°, y = 47°$
   (D) $x = 133°, y = 26°$

6. What type of triangle is illustrated?

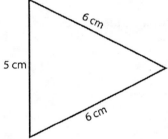

   (A) right
   (B) isosceles
   (C) equilateral
   (D) obtuse

7. Which side in $\triangle ABC$ is the longest side?

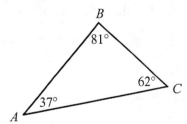

   (A) $\overline{AB}$
   (B) $\overline{AC}$
   (C) $\overline{BC}$
   (D) Cannot be determined

8. The incenter is also the center of the circle inscribed inside the triangle.

   (A) True
   (B) False - The incenter is outside.
   (C) False - The orthocenter is the center of the inscribed circle.
   (D) False - The incenter is the center of the circumscribed circle.

9. Which of the following lies on the middle of the hypotenuse?

   (A) Incenter
   (B) Centroid
   (C) Orthocenter
   (D) Circumcenter

10. Choose the theorem that proves the congruence.

   (A) SSS
   (B) SAS
   (C) AAS
   (D) The triangles are not congruent.

11. The meeting place of the altitudes is the

   (A) centroid.
   (B) incenter.
   (C) orthocenter.
   (D) circumcenter.

12. The angle bisectors meet at the

   (A) centroid.
   (B) incenter.
   (C) orthocenter.
   (D) circumcenter.

13. Which inequality theorem states that the sum of the measures of any two sides in a triangle must be greater than the measure of the third side in that triangle?

   (A) side-angle inequality
   (B) triangle inequality
   (C) exterior-angle inequality
   (D) hypotenuse-length theorem

14. Perpendicular bisectors of the sides meet at the

   (A) centroid.
   (B) incenter.
   (C) orthocenter.
   (D) circumcenter.

15. Which angle in the triangle below has the smallest measure?

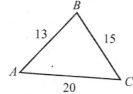

   (A) $\angle A$
   (B) $\angle B$
   (C) $\angle C$
   (D) All the angles are the same measure.

16. Choose the theorem that proves the congruence.

   (A) ASA
   (B) SAS
   (C) AAS
   (D) HL

Chapter 8
Quadrilaterals and Other Polygons

This chapter covers the following Georgia Performance Standards:

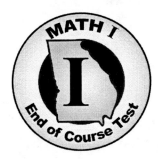

MM1G	Geometry	MM1G3a
		MM1G3d

8.1 Polygons

A **polygon** is a geometric figure with a finite number of sides in which each side intersects with two other sides only at their endpoints. A **convex polygon** is one in which every line segment connecting any pair of its points is entirely inside the figure's interior. If a polygon is not convex, it is a **concave polygon**. A concave polygon has an indentation and at least one interior angle that is greater than 180°. A **regular polygon** is one in which all the sides are the same measurement and all the angles are the same measurement (congruent).

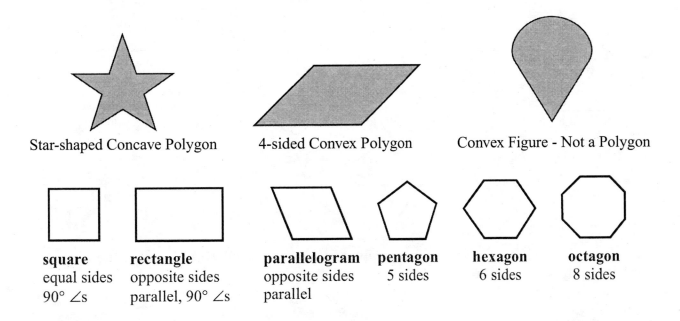

Star-shaped Concave Polygon   4-sided Convex Polygon   Convex Figure - Not a Polygon

square
equal sides
90° ∠s

rectangle
opposite sides
parallel, 90° ∠s

parallelogram
opposite sides
parallel

pentagon
5 sides

hexagon
6 sides

octagon
8 sides

8.2   Sum of Interior Angles of a Polygon

Given a polygon, you can find the sum of the measures of the interior angles using the following formula: Sum of the measures of the interior angles $= 180° (n - 2)$, where n is the number of sides of the polygon.

Example 1:   Find the sum of the measures of the interior angles of the following polygon:

Solution: The figure has 8 sides. Using the formula we have
$180° (8 - 2) = 180° (6) = 1080°$

Using the formula, $180° (n - 2)$, find the sum of the interior angles of the following figures.

1.

4.

7.

10.

2.

5.

8.

11.

3.

6.

9.

12.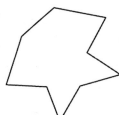

Find the measure of $\angle G$ in the regular polygons shown below. Remember that the sides of a regular polygon are equal.

13.

14.

15.

8.3   Exterior Angles of Polygons

The sum of the measures of the exterior angles of a convex polygon, one at each vertex, is $360°$.

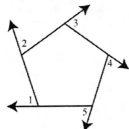

$$m\angle 1 + m\angle 2 + m\angle 3 + m\angle 4 + m\angle 5 = 360°.$$

In addition, for a regular n-gon, in which all angle measures are the same, the measure of each exterior angle is $\frac{1}{n} \times 360°$ or $\frac{360°}{n}$ where n is the number of sides.

Example 2:   Find the value of x.

Solution: The sum of the measures of the exterior angles is $360°$, so

$$
\begin{aligned}
90° + 90° + x + 2x + 3x &= 360° \\
180° + 6x &= 360° \\
6x &= 180° \\
x &= 30°
\end{aligned}
$$

Example 3:   One exterior angle of a regular convex n-gon measures $22.5°$. How many sides does it have?

Solution: The measure of each exterior angle of a convex n-gon is $\frac{360°}{n}$, where n is the number of sides. Therefore,

$$22.5° = \frac{360°}{n} \quad \text{so} \quad n = \frac{360°}{22.5°} = 16$$

The polygon has 16 sides.

Solve the following problems.

1. Find the measure of each exterior angle of a regular dodecagon (12 sides).

2. Find the measure of each exterior angle of a regular octagon.

3. Each exterior angle of a regular polygon equals $40°$. How many sides does the polygon have?

4. The exterior angles of a given polygon measure $45°$, $125°$, $2x°$, $3x°$, and $5x°$. Find the value of x and list the five angle measures.

5. If the measure of the exterior angles of a polygon are $x°$, $3x°$, $(x + 2)°$, $(5x - 4)°$, $30°$, and $42°$, find the value of x.

8.4   Quadrilaterals and Their Properties

A **quadrilateral** is a polygon with four sides. A **parallelogram** is a quadrilateral in which both pairs of opposite sides are parallel. The following properties of parallelograms are given without proof:

1. **Both pairs of opposite sides are parallel.**

2. **The opposite sides are congruent.**

3. **The opposite angles are congruent**

4. **Consecutive angles are supplementary.**

5. **The diagonals bisect each other**

A **rectangle** is a parallelogram with four right angles. It follows that a rectangle has all of the properties listed above, plus all four angles are 90°. In addition, the diagonals of a rectangle are congruent. A **rhombus** is a parallelogram with four congruent sides. A rhombus has all the properties of a parallelogram, but both pairs of opposite sides are congruent as well as parallel. Due to this fact, the diagonals of a rhombus each bisect a pair of opposite angles and the diagonals are perpendicular to each other. A **square** is a rhombus with four right angles. Therefore, a square has four congruent sides and four congruent angles (each 90°). As you can see, a square is also a quadrilateral, a parallelogram, a rectangle, and a rhombus. These properties plus the four right angles make it a square.

A **trapezoid** is a quadrilateral with only one pair of parallel sides. The parallel sides are called bases, and the other two sides are called legs.

8.5   More Properties of Quadrilaterals

There are two more kinds of quadrilaterals we need to know about: parallelograms and kites.

Examples:

1. **Parallelogram** – a figure with two sets of parallel sides. The opposite sides are equal in length and the opposite angles are congruent.

2. **Kite** – a figure with two pairs of congruent adjacent sides. One diagonal divides the kite into two congruent triangles. Two diagonals divide the kite into 4 triangles.

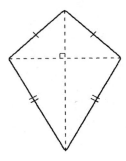

Use the following figure for questions 1–4.

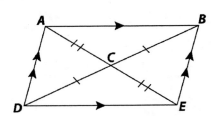

1. If $\overline{AB} = 6$, what is the measure of $\overline{DE}$?

2. If $\overline{AD} = 3$, what is the measure of $\overline{BE}$?

3. If $\overline{AC} = 5$, what is the measure of $\overline{AE}$?

4. If $\overline{BD} = 10$, what is the measure of $\overline{DC}$?

Use the following figure for questions 5–8.

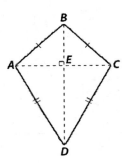

5. If $\overline{CD} = 5$, what is the measure of $\overline{AD}$?

6. If $\overline{AB} = 2$, what is the measure of $\overline{BC}$?

7. If $\overline{AE} = 3$, what is the measure of $\overline{AC}$?

8. If $\overline{CE} = 10$, what is the measure of $\overline{AE}$?

Example 4:   In the diagram below, $ABCD$ is a parallelogram. What is the length, in units, of $\overline{AC}$?

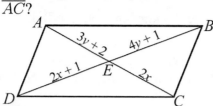

Step 1:   Since this is a parallelogram, $\overline{AE} = \overline{EC}$ and $\overline{DE} = \overline{EB}$. Knowing this we can set up two equations:
Equation 1: $3y + 2 = 2x$
Equation 2: $4y + 1 = 2x + 1$

Step 2:   These two equations are a system of equations. We will solve for y by subtracting equation 1 from equation 2. (Remember to subtract every term in equation 1.)

$$\begin{array}{rcl} 4y + 1 & = & 2x + 1 \\ -\,(3y + 2) & = & -2x \\ \hline y - 1 & = & 1 \\ y & = & 2 \end{array}$$

Step 3:   Now that we know $y = 2$, plug this value into either equation 1 or 2 to find the value of x.
$3y + 2 = 2x$
$3(2) + 2 = 2x$
$8 = 2x$
$4 = x$

Step 4:   Now that we know $y = 2$ and $x = 4$, we can find the length of $\overline{AC}$.
$\overline{AC} = 3y + 2 + 2x = 3(2) + 2 + 2(4) = 6 + 2 + 8 = 16$
$\overline{AC} = 16$

In the diagram below, $ABCD$ is a rectangle. Use the figure for questions 1–6.

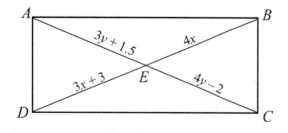

1. What is the value of x?

2. What is the value of y?

3. What is the measure of $\overline{AE}$?

4. What is the measure of $\overline{EB}$?

5. What is the measure of $\overline{AC}$?

6. What is the measure of $\overline{DB}$?

Chapter 8 Review

1. What is the sum of the measures of the interior angles in the figure below?

2. What is the value of one exterior angle of a regular decagon (10 sides)?

3. What is the sum of the exterior angles of a 28-gon?

Use the figure below to answer problems 4–6.

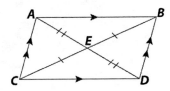

4. True or false. $\triangle ACD \cong \triangle DBA$

5. If $\overline{AC} = 2$, what is the measure of $\overline{BD}$?

6. If $\overline{CE} = 7$, what is the measure of $\overline{CB}$?

Use the figure below to answer problems 7–8.

7. True or false. $\triangle ABE \cong \triangle CBE$

8. If $\overline{BC} = 6$, what is the measure of $\overline{AB}$?

Chapter 8 Test

1. What is the sum of the measures of the interior angles of a polygon with 7 sides?

    (A) 180°
    (B) 360°
    (C) 540°
    (D) 900°

2. Which of the following figures is a parallelogram?

    (A)

    (B)

    (C)

    (D)

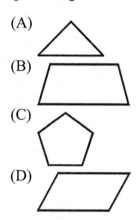

3. Which item below is not a polygon?

    (A) triangle
    (B) heptagon
    (C) octagon
    (D) circle

4. Find the values of x and y in the figure.

    (A) 55°, 125°
    (B) 90°, 90°
    (C) 55°, 90°
    (D) 90°, 125°

5. If a regular octagon has eight sides, what is the measure of each exterior angle?

    (A) 15°
    (B) 30°
    (C) 45°
    (D) 360°

6. What is the name of the polygon below?

    (A) quadrilateral
    (B) pentagon
    (C) hexagon
    (D) octagon

7. A square is

    (A) a quadrilateral.
    (B) parallelogram.
    (C) both.
    (D) neither.

8. A parallelogram with four congruent sides is

    (A) a rhombus.
    (B) a rectangle.
    (C) a square.
    (D) both A and C.

9. What is the sum of the interior angles of an octagon?

    (A) 1440°
    (B) 1080°
    (C) 1260°
    (D) 360°

10. What is the measure of angle B?

(A) 36°

(B) 90°

(C) 108°

(D) 144°

11. Find the value of x in the figure.

(A) 45°

(B) 51.4°

(C) 90°

(D) 360°

12. Lem wanted to cut a rusted stop sign into quarters in his shop class. Which of the following statements is factual?

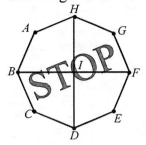

(A) $m\angle ABF = 67.5°$

(B) $m\angle CBF + m\angle BAH = 135°$

(C) $m\angle HGF = 67.5°$

(D) $m\angle BAH - m\angle HIF = 35°$

Chapter 9
Logic and Geometric Proofs

This chapter covers the following Georgia Performance Standards:

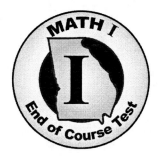

MM1G	Geometry	MM1G2a
		MM1G2b

9.1 Mathematical Reasoning/Logic

The Georgia mathematics curriculum calls for skill development in mathematical **reasoning** or **logic**. The ability to use logic is an important skill for solving math problems, but it can also be helpful in real-life situations. For example, if you need to get to Park Street, and the Park Street bus always comes to the bus stop at 3 PM, then you know that you need to get to the bus stop by at least 3 PM. This is a real-life example of using logic, which many people would call "common sense."

There are many different types of statements which are commonly used to describe mathematical principles. However, using the rules of logic, the truth of any mathematical statement must be evaluated. Below is a list of tools used in logic to evaluate mathematical statements.

Logic is the discipline that studies valid reasoning. There are many forms of valid arguments, but we will review just a few here.

A **proposition** is usually a declarative sentence which may be true or false.

An **argument** is a set of two or more related propositions, called **premises**, that provide support for another proposition, called the **conclusion**.

Deductive reasoning is an argument which begins with general premises and proceeds to a more specific conclusion. Most elementary mathematical problems use deductive reasoning.

Inductive reasoning is an argument in which the truth of its premises make it likely or probable that its conclusion is true.

9.2   Arguments

Most of logic deals with the evaluation of the validity of arguments. An argument is a group of statements that includes a conclusion and at least one premise. A premise is a statement that you know is true or at least you assume to be true. Then, you draw a conclusion based on what you know or believe is true in the premise(s). Consider the following example:

Premise 1

You know that

My socks are either in the top drawer or the bottom drawer.

Premise 2

You look to see that

My socks are not in the top drawer.

Conclusion

Therefore, without even looking, you know that

My socks are in the bottom drawer.

This argument is an example of deductive reasoning, where the conclusion is "deduced" from the premises and nothing else. In other words, if Premise 1 and Premise 2 are true, you don't even need to look in the bottom drawer to know that the conclusion is true.

9.3   Deductive and Inductive Arguments

In general, there are two types of logical arguments: **deductive** and **inductive**. Deductive arguments tend to move from general statements or theories to more specific conclusions. Inductive arguments tend to move from specific observations to general theories.

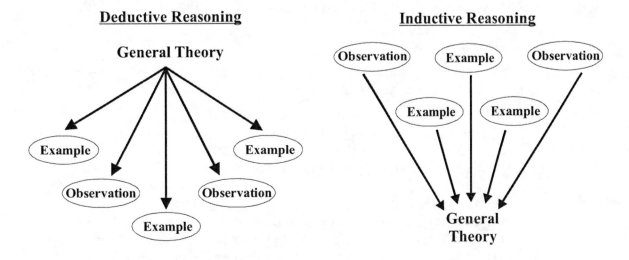

Deductive Reasoning

General Theory

Example
Observation
Example
Observation
Example

Inductive Reasoning

Observation   Example   Observation

Example   Example

General Theory

Compare the two examples below:

Deductive Argument		Inductive Argument	
Premise 1	All men are mortal.	Premise 1	The sun rose this morning.
Premise 2	Socrates is a man.	Premise 2	The sun rose yesterday morning.
Conclusion	Socrates is mortal.	Premise 3	The sun rose two days ago.
		Premise 4	The sun rose three days ago.
		Conclusion	The sun will rise tomorrow.

An inductive argument cannot be proved beyond a shadow of a doubt. For example, it's a pretty good bet that the sun will come up tomorrow, but the sun not coming up presents no logical contradiction. An inductive argument is a type of conjecture. A **conjecture** appears to be true, but has not been proven logically. It is an educated guess.

On the other hand, a deductive argument can have logical certainty, but it must be properly constructed. Consider the examples below.

True Conclusion from an Invalid Argument	**False Conclusion from a Valid Argument**
All men are mortal. Socrates is mortal. Therefore Socrates is a man.	All astronauts are men. Julia Roberts is an astronaut. Therefore, Julia Roberts is a man.
Even though the above conclusion is true, the argument is based on invalid logic. Both men and women are mortal. Therefore, Socrates could be a woman.	In this case, the conclusion is false because the premises are false. However, the logic of the argument is valid because *if* the premises were true, then the conclusion would be true.

A **counterexample** is an example in which the statement is true but the conclusion is false when we have assumed it to be true. If we said, "All cocker spaniels have blonde hair," then a counterexample would be a red-haired cocker spaniel. If we made the statement, "If a number is greater than 10, it is less than 20," we can easily think of a counterexample, like 35.

Example 1:   Which argument is valid?

If you speed on Hill Street, you will get a ticket.
If you get a ticket, you will pay a fine.

(A) I paid a fine, so I was speeding on Hill Street.
(B) I got a ticket, so I was speeding on Hill Street.
(C) I exceeded the speed limit on Hill Street, so I paid a fine.
(D) I did not speed in Hill Street, so I did not pay a fine.

Solution: C is valid.
A is incorrect. I could have paid a fine for another violation.
B is incorrect. I could have gotten a ticket for some other violation.
D is incorrect. I could have paid a fine for speeding somewhere else.

Example 2:   Assume the given proposition is true. Then determine if each statement is true or false.

Given: If a dog is thirsty, he will drink.

(A)	If a dog drinks, then he is thirsty.	T or F
(B)	If a dog is not thirsty, he will not drink.	T or F
(C)	If a dog will not drink, he is not thirsty.	T or F

Solution: A is false. He is not necessarily thirsty; he could just drink because other dogs are drinking or drink to show others his control of the water. This statement is the **converse** of the original. The converse of the statement "If A, then B" is "If B, then A."

B is false. The reasoning from A applies. This statement is the **inverse** of the original. The inverse of the statement "If A, then B" is "If not A, then not B."

C is true. It is the **contrapositive**, or the complete opposite of the original. The contrapositive says "If not B, then not A."

For numbers 1–5, what conclusion can be drawn from each proposition?

1. All squirrels are rodents. All rodents are mammals. Therefore,

2. All fractions are rational numbers. All rational numbers are real numbers. Therefore,

3. All squares are rectangles. All rectangles are parallelograms. All parallelograms are quadrilaterals. Therefore,

4. All Chevrolets are made by General Motors. All Luminas are Chevrolets. Therefore,

5. If a number is even and divisible by three, then it is divisible by six. Eighteen is divisible by six. Therefore,

For numbers 6–9, assume the given proposition is true. Then, determine if the statements following it are true or false.

All squares are rectangles.

6.	All rectangles are squares.	T or F
7.	All non-squares are non-rectangles.	T or F
8.	No squares are non-rectangles.	T or F
9.	All non-rectangles are non-squares.	T or F

9.4   Logic

A **conditional statement** is a type of logical statement that has two parts, a **hypothesis** and a **conclusion**. The statement is written in "if-then" form, where the "if" part contains the hypothesis and the "then" part contains the conclusion. Shorthand: $p \rightarrow q$. For example, let's start with the statement "Two lines intersect at exactly one point." We can rewrite this as a conditional statement in "if-then" form as follows:

If two lines intersect, then their intersection is at exactly one point.

hypothesis   conclusion

Conditional statements may be true or false. To show that a statement is false, you need only to provide a single **counterexample** which shows that the statement is not always true. To show that a statement is true, on the other hand, you must show that the conclusion is true for all occasions in which the hypothesis occurs. This is often much more difficult.

Example 3:   Provide a counterexample to show that the following conditional statement is false:
If $x^2 = 4$, then $x = 2$.

To find a counterexample, think: Could x equal something else? Yes, x could be -2. $(-2)^2 = 4$
Therefore, we have provided a counterexample to show that the conditional statement is false.

The **converse** of a conditional statement is an "if-then" statement written by switching the hypothesis and the conclusion. For example, for the conditional statement "If a figure is a quadrilateral, then it is a rectangle," the converse is "If a figure is a rectangle, then it is a quadrilateral." Shorthand: $q \rightarrow p$

The **inverse** of a conditional statement is written by negating the hypothesis and conclusion of the original "if-then" conditional statement. Negating means to change the meaning so it is the negative, or opposite, of its original meaning. The inverse of the conditional statement "If a figure is a quadrilateral, then it is a rectangle" is "If a figure is **not** a quadrilateral, then it is **not** a rectangle." Shorthand: $\sim p \rightarrow \sim q$

The **contrapositive** of a conditional statement is written by negating the converse. That is, switch the hypothesis and conclusion of the original statement, and make them both negative. The contrapositive of the conditional statement "If a figure is a quadrilateral, then it is a rectangle" is "If a figure is not a rectangle, then it is not a quadrilateral." Shorthand: $\sim q \rightarrow \sim p$

Example 4:   Given the conditional statement "If $m\angle F = 60°$, then $\angle F$ is acute." Write the converse, inverse and contrapositive.

   Step 1:   The converse (q → p) is constructed by switching the hypothesis and the conclusion: If $\angle F$ is acute, then $m\angle F = 60°$.

   Step 2:   The inverse (~p → ~q) is constructed by negating the original statement: If $m\angle F \neq 60°$, then $\angle F$ is not acute.

   Step 3:   The contrapositive (~q → ~p) is the negation of the converse: If $\angle F$ is not acute, then $m\angle F \neq 60°$.

Answer the following problems about geometry logic.

1. Rewrite the following as a conditional statement in "if-then" form: A number divisible by 8 is also divisible by 4.

2. Write the converse of the following conditional statement: If two circles have equal radii, then the circles are congruent.

3. Consider the conditional statement: If $x^4 = 81$, then $x = 3$. Is the statement true? Provide a counterexample if it is false.

4. Consider the statement: A line contains at least two points. Write as a conditional statement in "if-then" form, then write the converse, inverse, and contrapositive of the conditional statement.

5. "If a parallelogram has four congruent sides, then it is a rhombus." Write the converse, inverse, and contrapositive for the conditional statement. Which are true? Which are false?

6. "If a triangle has one right angle, then the acute angles are complementary." Write the converse, inverse, and contrapositive for the conditional statement. Indicate whether each is true or false. Can all the statements be either true or false? Explain.

7. "If a rectangle has four congruent sides, then it is a square." Write the contrapositive for the conditional statement and indicate whether it is true or false. Give a counterexample if it is false.

8. Show why a conditional statement and its inverse are always logically equivalent. Similarly, show why a statement's converse and inverse are logically equivalent.

Find the new if-then statement using the two statements below.

   p = It is raining.   q = We will not go outside.

9. Find the inverse. (~p → ~q)

10. Find the converse. (q → p)

11. Find the contrapositive. (~q → ~p)

Chapter 9 Review

For numbers 1–4, assume the given proposition is true. Then determine if the statements following it are true or false.

All whales are mammals.

1. All non-whales are non-mammals.

2. If a mammal lives in the sea, it is a whale.

3. All mammals are whales.

4. All non-mammals are non-whales.

For numbers 5–8, determine whether the situation is showing deductive or inductive logic.

5. A group of students were given three descriptions about a person's job. They were then told to decide what type of job title the person has.

6. When traveling in a car on a family vacation, I noticed that I could see the ocean to my left and palm trees to my right. I concluded that my family and I were going to the beach.

7. Sammy asked her friend, Amy, to give her a good reason to get a summer job. Amy gave Sammy four good reasons to get a job.

8. The neighbor's cars are in the driveway and all of the lights in the house are off, so they must be sleeping.

Solve the following problems.

9. "If a triangle is isosceles, then its base angles are congruent." Write the contrapositive for the conditional statement. Is the statement true or false? Is the contrapositive true or false? If false, give a counterexample.

10. "If the radius of a circle is doubled, then the area of the circle is increased by a factor of four." Write the converse, inverse, and contrapositive for the conditional statement. Indicate which ones are true or false.

11. "If today is Tuesday, then it is raining." Write the converse, inverse, and contrapositive for the conditional statement. Could the statements be true? Give a counterexample to prove each statement false.

For the next three questions, find the new if-then statement about the two statements below. Then determine if the new statement is true or false. If it is false, give an explanation.

p = A figure has two sets of parallel lines. q = The figure is a parallelogram.

12. Find the inverse. ($\sim p \rightarrow \sim q$)

13. Find the converse. ($q \rightarrow p$)

14. Find the contrapositive. ($\sim q \rightarrow \sim p$)

Chapter 9 Test

For 1–3, chose which argument is valid.

1. If I oversleep, I miss breakfast. If I miss breakfast, I cannot concentrate in class. If I do not concentrate in class, I make bad grades.

   (A) I made bad grades today, so I missed breakfast.
   (B) I made good grades today, so I got up on time.
   (C) I could not concentrate in class today, so I overslept.
   (D) I had no breakfast today, so I overslept.

2. If I do not maintain my car regularly, it will develop problems. If my car develops problems, it will not be safe to drive. If my car is not safe to drive, I cannot take a trip in it.

   (A) If my car develops problems, I did not maintain it regularly.
   (B) I took a trip in my car, so I maintained it regularly.
   (C) If I maintain my car regularly, it will not develop problems.
   (D) If my car is safe to drive, it will not develop problems.

3. If two triangles have all corresponding sides and all corresponding angles congruent, then they are congruent triangles. If two triangles are congruent, then they are similar triangles.

   (A) Similar triangles have all sides and all angles congruent.
   (B) If two triangles are similar, then they are congruent.
   (C) If two triangles are not congruent, then they are not similar.
   (D) If two triangles have all corresponding sides and angles congruent, then they are similar triangles.

4. Cynthia is asked to list five duties of the President. What type of logic is Cynthia using?

   (A) mathematical reasoning
   (B) inductive reasoning
   (C) intuitive reasoning
   (D) deductive reasoning

Answer the next three questions about the two statements below.

$$p = \text{A figure has four sides.} \qquad q = \text{It is a square.}$$

5. What is the inverse, and is it true?

   (A) $\sim p \rightarrow \sim q$, yes
   (B) $\sim p \rightarrow \sim q$, no
   (C) $p \rightarrow q$, yes
   (D) $q \rightarrow p$, no

6. What is the converse, and is it true?

   (A) $q \rightarrow p$, yes
   (B) $\sim q \rightarrow \sim p$, yes
   (C) $\sim p \rightarrow \sim q$, no
   (D) $q \rightarrow p$, no

7. What is the contrapositive, and is it true?

   (A) $\sim q \rightarrow \sim p$, yes
   (B) $p \rightarrow q$, yes
   (C) $q \rightarrow p$, no
   (D) $\sim q \rightarrow \sim p$, no

Chapter 10
Permutations and Combinations

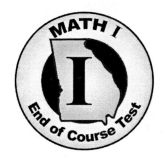

This chapter covers the following Georgia Performance Standards:

MM1D	Data Analysis and Probability	MM1D1b

10.1   Permutations

A **permutation** is an arrangement of items in a specific order. The formula $_nP_r = \dfrac{n!}{(n-r)!}$ is the formula for permutations. The number n is the number you have to choose from, and r is the number of objects you want to arrange. If a problem asks how many ways can you arrange 6 books on a bookshelf, it is asking how many permutations there are for 6 items.

Example 1:   Ron has 4 items: a model airplane, a trophy, an autographed football, and a toy sports car. How many ways can he arrange the 4 items on a shelf?

Solution: Ron has 4 choices for the first item on the shelf. He then has 3 choices left for the second item. After choosing the second item, he has 2 choices left for the third item and only one choice for the last item. The diagram below shows the permutations for arranging the 4 items on a shelf if he chose to put the trophy first.

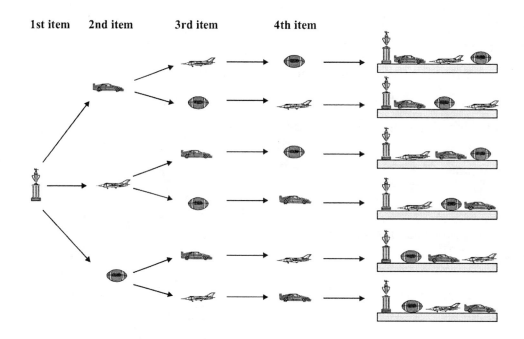

Copyright ©American Book Company

Count the number of permutations if Ron chooses the trophy as the first item. There are 6 permutations. Next, you could construct a pyramid of permutations choosing the model car first. That pyramid would also have 6 permutations. Then, you could construct a pyramid choosing the airplane first. Finally, you could construct a pyramid choosing the football first. You would then have a total of 4 pyramids each having 6 permutations. The total number of permutations is $6 \times 4 = 24$. There are 24 ways to arrange the 4 items on a bookshelf.

You probably don't want to draw pyramids for every permutation problem. What if you want to know the permutations for arranging 30 objects? Fortunately, mathematicians have come up with a formula for calculating permutations.

For the above problem, Ron has 4 items to arrange. Therefore, multiply $4 \times 3 \times 2 \times 1 = 24$. Another way of expressing this calculation is $4!$, stated as 4 factorial. $4! = 4 \times 3 \times 2 \times 1$.

Note: To find the permutation using the formula, $_nP_r = \dfrac{n!}{(n-r)!}$. n is the number you have to choose from, 4, and r is the number of objects you want to arrange, 4.

$$_4P_4 = \frac{4!}{(4-4)!} = \frac{4!}{0!} = \frac{4!}{1} = 4! = 4 \times 3 \times 2 \times 1 = 24$$

(Remember that $0! = 1$.)

Example 2:   How many ways can you line up 6 students?

Solution: The number of permutations for 6 students $= 6! = 6 \times 5 \times 4 \times 3 \times 2 \times 1 = 720$. There are 6 choices for the first position, 5 for the second position, 4 for the third, 3 for the fourth, 2 for the fifth, and 1 for the sixth.

Note: The numbers of permutation for 6 students $= {_6P_6} = \dfrac{6!}{(6-6)!} = \dfrac{6!}{0!} = \dfrac{6!}{1} = 6! = 720$.

Example 3:   Shelley and her mom, dad, and brother are having cake for her birthday. Since it is Shelley's birthday she gets a piece first. How many ways are there to pass out the pieces of cake?

Solution: Since Shelley gets the first piece, the first spot is fixed. The second, third, and fourth spots are not fixed and anyone left can be in one of the three spots.

Spot	1	2	3	4
Choices of people	1	3	2	1

Now, multiply the choices together, $1 \times 3 \times 2 \times 1 = 6$ ways to pass out cake.

Note: You can also use the permutation formula, but fix the first spot and just arrange the last three spots.

$$1 \times {_3P_3} = \frac{3!}{(3-3)!} = \frac{3!}{0!} = \frac{3!}{1} = 3! = 3 \times 2 \times 1 = 6.$$

Work the following permutation problems.

1. How many ways can you arrange 5 books on a bookshelf?

2. Myra has 6 novels to arrange on a book shelf. How many ways can she arrange the novels?

3. Seven sprinters signed up for the 100 meter dash. How many ways can the seven sprinters line up on the starting line?

4. Keri wants an ice cream cone with one scoop of chocolate, one scoop of vanilla, and one scoop of strawberry. How many ways can the scoops be arranged on the cone?

5. How many ways can you arrange the letters A, B, C, and D?

6. At Sam's party, the DJ has four song requests. In how many different orders can he play the 4 songs?

7. Yvette has 5 comic books. How many different ways can she stack the comic books?

8. Sandra's couch can hold three people. How many ways can she and her two friends sit on the couch?

9. How many ways can you arrange the numbers 2, 3, 5?

10. At a busy family restaurant, four tables open up at the same time. How many different ways can the hostess seat the next four families waiting to be seated?

11. How many ways can you arrange the numbers 1, 2, 3, 4, 5, 6, 7, 8, 9, 10 and always have 3 at position 1 and 10 at position 5?

10.2 More Permutations

The formula $_nP_r = \dfrac{n!}{(n-r)!}$ can also be used if you are trying to arrange a specific number of objects, but have more than you want to arrange. The number n is the number you have to choose from, and r is the number of objects you want to arrange.

Example 4: If there are 6 students, how many ways can you line up any 4 of them?

Step 1: Find all the variables. The formula is $_nP_r = \dfrac{n!}{(n-r)!}$.

n (the number you have to choose from) = 6
r (the number of objects you want to arrange) = 4

Step 2: Plug the variables into the formula.

$$_6P_4 = \frac{6!}{(6-4)!} = \frac{6!}{2!} = \frac{6 \times 5 \times 4 \times 3 \times \cancel{2} \times \cancel{1}}{\cancel{2} \times \cancel{1}} = 6 \times 5 \times 4 \times 3 = 360$$

There are 360 ways to line up 4 of the 6 students.

Find the number of permutations for each of the problems below.

1. How many ways can you arrange 4 out of 8 books on a shelf?

2. How many 3 digit numbers can be made using the numbers 2, 3, 5, 8, and 9?

3. How many ways can you line up 4 students out of a class of 20?

4. Kim worked in the linen department of a store. Eight new colors of towels came in. Her job was to line up the new towels on a long shelf. How many ways can she arrange the 8 colors?

5. Terry's CD player holds 5 CDs. Terry owns 12 CDs. How many different ways can he arrange his CDs in the CD player?

6. Erik has 11 shirts he wears to school. How many ways can he choose a different shirt to wear on Monday, Tuesday, Wednesday, Thursday, and Friday?

7. Deb has a box of 12 markers. The art teacher told her to choose three markers and line them up on her desk. How many ways can she line up 3 markers from the 12?

8. Jeff went into an ice cream store serving 32 flavors of ice cream. He wanted a cone with two different flavors. How many ways could he order 2 scoops of ice cream, one on top of the other?

9. In how many ways can you arrange any 3 letters from the 26 letters in the alphabet?

10.3   Combinations

In a **permutation**, objects are arranged in a particular order. In a **combination**, the order does not matter. In a **permutation**, if someone picked two letters of the alphabet, **k, m** and **m, k** would be considered 2 different permutations. In a **combination**, **k, m** and **m, k** would be the same combination. A different order does not make a new combination. The formula for combinations is $_nC_r = \binom{n}{r} = \dfrac{n!}{(n-r)!r!}$ where n is the total number of objects you have to choose from and r is the number that you choose to arrange.

Example 5:   How many combinations of 3 letters from the set {a, b, c, d, e} are there?

Step 1:   Find the **permutation** of 3 out of 5 objects.

Step 2:   Divide by the permutation of the **number of objects** to be chosen from the total (3). This step eliminates the duplicates in finding the permutations.

$$\frac{5 \times \overset{2}{\cancel{4}} \times \cancel{3}}{\cancel{3} \times \cancel{2} \times 1} = 10$$

Step 3:   Cancel common factors and simplify.

Note: Using the formula, find all your variables. The formula is $_nC_r = \binom{n}{r} = \dfrac{n!}{(n-r)!r!}$.

n (the number you choose from) = 5, r (the number of objects arranged) = 3.

$$_5C_3 = \binom{5}{3} = \frac{5!}{(5-3)!3!} = \frac{5!}{2!3!} = \frac{5 \times 4 \times \cancel{3} \times \cancel{2} \times \cancel{1}}{(2 \times 1)(\cancel{3} \times \cancel{2} \times \cancel{1})} = \frac{20}{2} = 10$$

There can be 10 combinations of three letters from the set {a, b, c, d, e}.

Find the number of combinations for each problem below.

1. How many combinations of 4 numbers can be made from the set of numbers {2, 4, 6, 7, 8, 9}?

2. Johnston Middle School wants to choose 3 students at random from the 7th grade to take an opinion poll. There are 124 seventh graders in the school. How many different groups of 3 students could be chosen? (Use a calculator for this one.)

3. How many combinations of 3 students can be made from a class of 20?

4. Fashion Ware catalog has a sweater that comes in 8 colors. How many combinations of 2 different colors does a shopper have to choose from?

5. Angelo's Pizza offers 10 different pizza toppings. How many different combinations can be made of pizzas with four toppings?

6. How many different combinations of 5 flavors of jelly beans can you make from a store that sells 25 different flavors of jelly beans?

7. The track team is running the relay race in a competition this Saturday. There are 14 members of the track team. The relay race requires 4 runners. How many combinations of 4 runners can be formed from the track team?

8. Kerri got to pick 2 prizes from a grab bag containing 12 prizes. How many combinations of 2 prizes are possible?

10.4   More Combinations

Another kind of combination involves selection from several categories.

Example 6:   At Joe's Deli, you can choose from 4 kinds of bread, 5 meats, and 3 cheeses when you order a sandwich. How many different sandwiches can be made with Joe's choices for breads, meats, and cheeses if you choose 1 kind of bread, 1 meat, and 1 cheese for each sandwich?

JOE'S SANDWICHES

Breads	**Meats**	**Cheeses**
White	Roast beef	Swiss
Pumpernickel	Corned beef	American
Light rye	Pastrami	Mozzarella
Whole wheat	Roast chicken	
	Roast turkey	

Solution: Multiply the number of choices in each category. There are 4 breads, 5 meats, and 3 cheeses, so $4 \times 5 \times 3 = 60$. There are 60 combinations of sandwiches.

Find the number of combinations that can be made in each of the problems below.

1. Angie has 4 pairs of shorts, 6 shirts, and 2 pairs of tennis shoes. How many different outfit combinations can be made with Angie's clothes?

2. Raymond has 7 baseball caps, 2 jackets, 10 pairs of jeans, and 2 pairs of sneakers. How many combinations of the 4 items can he make?

3. Claire has 6 kinds of lipstick, 4 eye shadows, 2 kinds of lip liner, and 2 mascaras. How many combinations can she use to make up her face?

4. Clarence's dad is ordering a new truck. He has a choice of 5 exterior colors, 3 interior colors, 2 kinds of seats, and 3 sound systems. How many combinations does he have to pick from?

5. A fast food restaurant has 8 kinds of sandwiches, 3 kinds of French fries, and 5 kinds of soft drinks. How many combinations of meals could you order if you ordered a sandwich, fries, and a drink?

6. In summer camp, Tyrone can choose from 4 outdoor activities, 3 indoor activities, and 3 water sports. He has to choose one of each. How many combinations of activities can he choose?

7. Jackie won a contest at school and gets to choose one pencil and one pen from the school store and an ice cream from the lunch room. There are 5 colors of pencils, 3 colors of pens, and 4 kinds of ice cream. How many combinations of prize packages can she choose?

Chapter 10 Review

Answer the following problems.

1. Daniel has 7 trophies he has won playing soccer. How many different ways can he arrange them in a row on his bookshelf?

2. Missy has 12 colors of nail polish. She wears 1 color each day, 7 different colors a week. How many combinations of 7 colors can she make before she has to repeat the same 7 colors in a week?

3. Eileen has a collection of 12 antique hats. She plans to donate 5 of the hats to a museum. How many combinations of hat are possible for her donation?

4. Julia has 5 porcelain dolls. How many ways can she arrange 3 of the dolls on a display shelf?

5. Ms. Randal has 10 students. Every day she randomly draws names of 2 students out of a bag to turn in their homework for a test grade. How many combinations of 2 students can she draw?

6. In the lunch line, students can choose 1 out of 3 meats, 1 out of 4 vegetables, 1 out of 3 desserts, and 1 out of 5 drinks. How many lunch combinations are there?

7. Andrea has 7 teddy bears in a row on a shelf in her room. How many ways can she arrange the bears in a row on her shelf?

8. Adrianna has 4 hats, 8 shirts, and 9 pairs of pants. Choosing one of each, how many different clothes combinations can she make?

9. The buffet line offers 5 kinds of meat, 3 different salads, a choice of 4 desserts, and 5 different drinks. If you choose one food from each category, from how many combinations would you have to choose?

10. How many pairs of students can Mrs. Smith choose to go to the library if she has 20 students in her class?

Chapter 10 Test

1. How many ways can you arrange six books on a bookshelf?

   (A) 6
   (B) 12
   (C) 36
   (D) 720

2. Katie and Kayla are having a party. They each invite four friends. How many ways could they set around a rectangular table if Katie is always sitting at the left end of the table and Kayla is always at the right end of the table?

Katie   Kayla

   (A) 80
   (B) 6400
   (C) 40,320
   (D) 50,000

3. How many ways can you line up any four students out of a class of twenty?

   (A) 116,280
   (B) 118,260
   (C) 260,118
   (D) 112,280

4. How many combinations of 3 numbers can be made from the set of numbers $\{2, 3, 5, 6, 7, 9\}$?

   (A) 5
   (B) 10
   (C) 15
   (D) 20

5. Savannah is going to the school dance. She has three dresses, six necklaces, and eight pairs of earrings. How many different outfits does she have to chose from if she chooses only one dress, one necklace, and one pair of earrings?

   (A) 17
   (B) 144
   (C) 48
   (D) 12

6. Refer to question 5 to answer the following problem.
   Savannah finds two more necklaces that she can chose from. How many difference outfits does she have to chose from now?

   (A) 192
   (B) 144
   (C) 132
   (D) 121

Chapter 11
Probability

This chapter covers the following Georgia Performance Standards:

MM1D	Data Analysis and Probability	MM1D1a
		MM1D2a
		MM1D2b
		MM1D2c
		MM1D2d

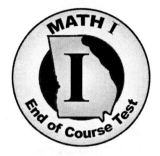

11.1   Probability Terms

Probability - the branch of mathematics that calculates the chance something will or will not happen.

Independent Events - the outcome of one event does not influence the outcome of the second event.

Dependent Events - the outcome of one event does influence the outcome of the second event.

Mutually Exclusive Events - two events that have no outcomes in common. These events cannot occur at the same time.

Conditional Probability - probability that a second event will happen, given that the first event has already occurred.

Equally Likely Outcomes - all outcomes of the event have the same chance of occurring.

Expected Value - the mean of a random variable.

Population - an entire group or collection about which we wish to draw conclusions.

Census - a count of the entire population

Sample - units selected to study from the population.

Sample Space - the set of all possible outcomes.

$P(A)$ - notation used to mean the probability of outcome 'A' occurring.

$P(A^c)$ (**Complement**) - notation used to mean the probability that outcome 'A' does not occur.

$P(A^c) = 1 - P(A)$

11.2 Probability

Probability is the chance something will happen. Probability is most often expressed as a fraction, a decimal, a percent, or can also be written out in words.

Example 1: Billy has 3 red marbles, 5 white marbles, and 4 blue marbles on the floor. His cat comes along and bats one marble under the chair. What is the **probability** it is a red marble?

Step 1: The number of red marbles will be the top number of the fraction. $\longrightarrow \dfrac{3}{12}$

Step 2: The total number of marbles is the bottom number of the fraction.

The answer may be expressed in lowest terms. $\dfrac{3}{12} = \dfrac{1}{4}$.

Expressed as a decimal, $\frac{1}{4} = 0.25$, as a percent, $\frac{1}{4} = 25\%$, and written out in words $\frac{1}{4}$ is one out of four.

Example 2: Determine the probability that the pointer will stop on a shaded wedge or the number 1.

Step 1: Count the number of possible wedges that the spinner can stop on to satisfy the above problem. There are 5 wedges that satisfy it (4 shaded wedges and one number 1). The top number of the fraction is 5.

Step 2: Count the total number of wedges, 7. The bottom number of the fraction is 7. The answer is $\frac{5}{7}$ or **five out of seven.**

Example 3: Refer to the spinner above. If the pointer stops on the number 7, what is the probability that it will **not** stop on 7 the next time?

Step 1: Ignore the information that the pointer stopped on 7 the previous spin. The probability of the next spin does not depend on the outcome of the previous spin. Simply find the probability that the spinner will **not** stop on 7. Remember, if P is the probability of an event occurring, $1 - P$ is the probability of an event **not** occurring (it is the complement). In this example, the probability of the spinner landing on 7 is $\frac{1}{7}$.

Step 2: The probability that the spinner will not stop on 7 is $1 - \frac{1}{7}$ which equals $\frac{6}{7}$. The answer is $\frac{6}{7}$ or **six out of seven.**

Find the probability of the following problems. Express the answer as a percent.

1. A computer chooses a random number between 1 and 50. What is the probability that you will guess the same number that the computer chose in 1 try?

2. There are 24 candy-coated chocolate pieces in a bag. Eight have defects in the coating that can be seen only with close inspection. What is the probability of pulling out a defective piece without looking?

3. Seven sisters have to choose which day each will wash the dishes. They put equal-sized pieces of paper in a hat, each labeled with a day of the week. What is the probability that the first sister who draws will choose a weekend day?

4. For his garden, Clay has a mixture of 12 white corn seeds, 24 yellow corn seeds, and 16 bicolor corn seeds. If he reaches for a seed without looking, what is the probability that Clay will plant a bicolor corn seed first?

5. Mom just got a new department store credit card in the mail. What is the probability that the last digit is an odd number?

6. Alex has a paper bag of cookies that holds 8 chocolate chip, 4 peanut butter, 6 butterscotch chip, and 12 ginger. Without looking, his friend John reaches in the bag for a cookie. What is the probability that the cookie is peanut butter?

7. An umpire at a little league baseball game has 14 balls in his pockets. Five of the balls are brand A, 6 are brand B, and 3 are brand C. What is the probability that the next ball he throws to the pitcher is a brand C ball?

8. What is the probability that the spinner's arrow will land on an even number?

9. The spinner in the problem above stopped on a shaded wedge on the first spin and stopped on the number 2 on the second spin. What is the probability that it will not stop on a shaded wedge or on the 2 on the third spin?

10. A company is offering 1 grand prize, 3 second place prizes, and 25 third place prizes based on a random drawing of contest entries. If your entry is one of the 500 total entries, what is the probability you will win a third place prize?

11. In the contest problem above, what is the probability that you will win the grand prize or a second place prize?

12. A box of a dozen doughnuts has 3 lemon cream-filled, 5 chocolate cream-filled, and 4 vanilla cream-filled. If the doughnuts look identical, what is the probability of picking a lemon cream-filled?

11.3   More Probability

Example 4:   You have a cube with one number, 1, 2, 3, 4, 5, and 6 painted on each face of the cube. What is the probability that if you throw the cube 3 times, you will get the number 2 each time?

If you roll the cube once, you have a 1 in 6 chance of getting the number 2. If you roll the cube a second time, you again have a 1 in 6 chance of getting the number 2. If you roll the cube a third time, you again have a 1 in 6 chance of getting the number 2. The probability of rolling the number 2 three times in a row is:

$$\frac{1}{6} \times \frac{1}{6} \times \frac{1}{6} = \frac{1}{216}$$

Find the probability that each of the following events will occur.

There are 10 balls in a box, each with a different digit on it: 0, 1, 2, 3, 4, 5, 6, 7, 8, & 9. A ball is chosen at random and then put back in the box.

1. What is the probability that if you pick out a ball 3 times, you will get number 7 each time?

2. What is the probability you will pick a ball with 5, then 9, and then 3?

3. What is the probability that if you pick out a ball 4 times, you will always get an odd number?

4. A couple has 4 children ages 9, 6, 4, and 1. What is the probability that they are all girls?

There are 26 letters in the alphabet, allowing a different letter to be on each of 26 cards. The cards are shuffled. After each card is chosen at random, it is put back in the stack of cards, and the cards are shuffled again.

5. What is the probability that when you pick 3 cards, you would draw first a "y", then an "e", and then an "s"?

6. What is the probability that you would draw 4 cards and get the letter "z" each time?

7. What is the probability that you draw twice and get a letter in the word "random" both times?

8. If you flip a coin 3 times, what is the probability you will get heads every time?

9. Marie is clueless about 4 of her multiple-choice answers. The possible answers are A, B, C, D, E, or F. What is the probability that she will guess all four answers correctly?

11.4   Tree Diagrams

Drawing a tree diagram is another method of determining the probability of events occurring.

Example 5:   If you toss two six-sided numbered cubes that have 1, 2, 3, 4, 5, or 6 on each side, what is the probability you will get two cubes that add up to 9? One way to determine the probability is to make a tree diagram.

Cube 1	Cube 2	Cube 1 plus Cube 2
1	1	2
	2	3
	3	4
	4	5
	5	6
	6	7
2	1	3
	2	4
	3	5
	4	6
	5	7
	6	8
3	1	4
	2	5
	3	6
	4	7
	5	8
	6	(9)
4	1	5
	2	6
	3	7
	4	8
	5	(9)
	6	10
5	1	6
	2	7
	3	8
	4	(9)
	5	10
	6	11
6	1	7
	2	8
	3	(9)
	4	10
	5	11
	6	12

Alternative method

Write down all of the numbers on both cubes which would add up to 9.

Cube 1	Cube 2
4	5
5	4
6	3
3	6

Numerator = 4 combinations

For denominator: Multiply the number of sides on one cube times the number of sides on the other cube.

$6 \times 6 = 36$

Numerator:
Denominator: $\dfrac{4}{36} = \dfrac{1}{9}$

There are 36 possible ways the cubes could land. Out of those 36 ways, the two cubes add up to 9 only 4 times. The probability you will get two cubes that add up to 9 is $\dfrac{4}{36}$ or $\dfrac{1}{9}$.

Read each of the problems below. Then answer the questions.

1. Jake has a spinner. The spinner is divided into eight equal regions numbered 1–8. In two spins, what is the probability that the numbers added together will equal 12?

2. Charlie and Libby each spin one spinner one time. The spinner is divided into 5 equal regions numbered 1–5. What is the probability that these two spins added together would equal 7?

3. Gail spins a spinner twice. The spinner is divided into 9 equal regions numbered 1–9. In two spins, what is the probability that the difference between the two numbers will equal 4?

4. Diedra throws two 10-sided numbered polyhedrons. What is the probability that the difference between the two numbers will equal 7?

5. Cameron throws two six-sided numbered cubes. What is the probability that the difference between the two numbers will equal 3?

6. Tesla spins one spinner twice. The spinner is divided into 11 equal regions numbered 1–11. What is the probability that the two numbers added together will equal 11?

7. Samantha decides to roll two five-sided numbered prisms. What is the probability that the two numbers added together will equal 4?

8. Mary Ellen spins a spinner twice. The spinner is divided into 7 equal regions numbered 1–7. What is the probability that the product of the two numbers equals 10?

9. Conner decides to roll two six-sided numbered cubes. What is the probability that the product of the two numbers equals 4?

10. Tabitha spins one spinner twice. The spinner is divided into 9 equal regions numbered 1–9. What is the probability that the sum of the two numbers equals 10?

11. Darnell decides to roll two 15-sided numbered polyhedrons. What is the probability that the difference between the two numbers is 13?

12. Inez spins one spinner twice. The spinner is divided into 12 equal regions numbered 1–12. What is the probability that the sum of two numbers equals 10?

13. Gina spins one spinner twice. The spinner is divided into 8 equal regions numbered 1–8. What is the probability that the two numbers added together equals 9?

14. Celia rolls two six-sided numbered cubes. What is the probability that the difference between the two numbers is 2?

15. Brett spins one spinner twice. The spinner is divided into 4 equal regions numbered 1–4. What is the probability that the difference between the two numbers will be 3?

11.5   Independent and Dependent Events

In mathematics, the outcome of an event may or may not influence the outcome of a second event. If the outcome of one event does not influence the outcome of the second event, these events are **independent**. However, if one event has an influence on the second event, the events are **dependent**. When someone needs to determine the probability of two events occurring, he or she will need to use an equation. These equations will change depending on whether the events are independent or dependent in relation to each other. When finding the probability of two **independent** events, multiply the probability of each favorable outcome together. Independent events use the **multiplication rule**. The multiplication rule for independent events is $P(A \text{ and } B) = P(A) P(B)$.

Example 6:   One bag of marbles contains 1 white, 1 yellow, 2 blue, and 3 orange marbles. A second bag of marbles contains 2 white, 3 yellow, 1 blue, and 2 orange marbles. What is the probability of drawing a blue marble from each bag?

Solution: Probability of favorable outcomes

Bag 1 $P(A)$: $\dfrac{2}{7}$

Bag 2 $P(B)$: $\dfrac{1}{8}$

Probability of a blue marble from each bag $= P(A \text{ and } B)$

$$P(A \text{ and } B) = P(A) P(B) = \frac{2}{7} \times \frac{1}{8} = \frac{2}{56} = \frac{1}{28}$$

In order to find the probability of two **dependent** events, you will need to use a different set of rules. For the first event, you must divide the number of favorable outcomes by the number of possible outcomes. For the second event, you must subtract one from the number of favorable outcomes **only if** the favorable outcome is the **same**. However, you must subtract one from the number of total possible outcomes. Finally, you must multiply the probability for event one by the probability for event two.

Example 7:   One bag of marbles contains 3 red, 4 green, 7 black, and 2 yellow marbles. What is the probability of drawing a green marble, removing it from the bag, and then drawing another green marble without looking?

	Favorable Outcomes	Total Possible Outcomes
Draw 1	4	16
Draw 2	3	15
Draw 1 × Draw 2	12	240

Answer: $\dfrac{12}{240}$ or $\dfrac{1}{20}$

Example 8:   Using the same bag of marbles, what is the probability of drawing a red marble and then drawing a black marble?

	Favorable Outcomes	Total Possible Outcomes
Draw 1	3	16
Draw 2	7	15
Draw 1 × Draw 2	21	240

Answer $\dfrac{21}{240}$ or $\dfrac{7}{80}$

Find the probability of the following problems. Express the answer as a fraction.

1. Prithi has two boxes. Box 1 contains 3 red, 2 silver, 4 gold, and 2 blue combs. She also has a second box containing 1 black and 1 clear brush. What is the probability that Prithi selects a red comb from box 1 and a black brush from box 2?

2. Steve Marduke has two spinners in front of him. The first one is numbered 1–6, and the second is numbered 1–3. If Steve spins each spinner once, what is the probability that the first spinner will show an odd number and the second spinner will show a "1"?

3. Carrie McCallister flips a coin twice and gets heads both times. What is the probability that Carrie will get tails the third time she flips the coin?

4. Artie Drake turns a spinner which is evenly divided into 11 sections numbered 1–11. On the first spin, Artie's pointer lands on "8". What is the probability that the spinner lands on an even number the second time he turns the spinner?

5. Leanne Davis plays a game with a street entertainer. In this game, a ball is placed under one of three coconut halves. The vendor shifts the coconut halves so quickly that Leanne can no longer tell which coconut half contains the ball. She selects one and misses. The entertainer then shifts all three around once more and asks Leanne to pick again. What is the probability that Leanne will select the coconut half containing the ball?

6. What is the probability that Jane Robelot reaches into a bag containing 1 daffodil and 2 gladiola bulbs and pulls out a daffodil bulb, and then reaches into a second bag containing 6 tulip, 3 lily, and 2 gladiola bulbs and pulls out a lily bulb?

7. Terrell casts his line into a pond containing 7 catfish, 8 bream, 3 trout, and 6 northern pike. He immediately catches a bream. What are the chances that Terrell will catch a second bream the next time he casts his line?

8. Gloria Quintero enters a contest in which the person who draws his or her initials out of a box containing all 26 letters of the alphabet wins the grand prize. Gloria reaches in, draws a "G", keeps it, then draws another letter. What is the probability that Gloria will next draw a "Q"?

9. Vince Macaluso is pulling two socks out of a washing machine in the dark. The washing machine contains three tan, one white, and two black socks. If Vince reaches in and pulls out the socks one at a time, what is the probability that he will pull out two tan socks on his first two tries?

10. John Salome has a bag containing 2 yellow plums, 2 red plums, and 3 purple plums. What is the probability that he reaches in without looking and pulls out a yellow plum and eats it, then reaches in again without looking and pulls out a red plum to eat?

11.6   Mutually Exclusive Events

Events are said to be **mutually exclusive** if they don't occur at the same time (no common outcome). This means the probability of event 1 (A) and event 2 (B) both occurring is zero, $P(A$ and $B) = 0$. Mutually exclusive events use the **addition rule**.

Let A and B be events. Let $P(A) = $ the probability of event A occurring and $P(B) = $ the probability of event B occurring.

The addition rule for mutually exclusive events is $P(A$ or $B) = P(A) + P(B)$.

The rule for sets that are **not** mutually exclusive is $P(A$ or $B) = P(A) + P(B) - P(A \cap B)$.

Example 9:   A pair of diced is rolled. What is the probability that the sum of the dice rolled is either a 7 or a 2?

Step 1:   Find the number of outcomes for rolling a 7.
There are six outcomes for rolling a seven:
$(1, 6), (6, 1), (2, 5), (5, 2), (3, 4), (4, 3)$

$$P(7) = \frac{\text{\# of outcomes for rolling a seven}}{\text{total number of outcomes}} = \frac{6}{36} = \frac{1}{6}$$

Step 2:   Then find the number of outcomes for rolling a 2.
There is one outcome for rolling a two:
$(1, 1)$

$$P(2) = \frac{\text{\# of outcomes for rolling a two}}{\text{total number of outcomes}} = \frac{1}{36}$$

Step 3:   Since the sum of the dice cannot be seven and two (it must be one or the other), then the events are mutually exclusive. Use the formula $P(x$ or $y) = P(x) + P(y)$ for mutually exclusive events to find the probability of either rolling a sum of 7 or a sum of 2.
Let $x = $ rolling a sum of 7 and $y = $ rolling a sum of 2

$$P(7 \text{ or } 2) = P(7) + P(2) = \tfrac{1}{6} + \tfrac{1}{36} = \tfrac{7}{36}$$

The probability of rolling the sum of either a 7 or a 2 is $\frac{7}{36}$.

Find the probability of each event.

1. A pair of dice is rolled. What is the probability that the sum of the dice rolled is either an 11 or a 5?

2. A pair of dice is rolled. What is the probability that the sum of the dice rolled is either a 9 or a 3?

3. A pair of dice is rolled. What is the probability that the sum of the dice rolled is either a 6 or a 12?

4. Russell has a bag of 100 marbles. 50 are red and 50 are black. What is the probability of
   (A) picking 2 red marbles if the first one is NOT replaced?
   (B) picking two black and two red marbles without replacement?

11.7 Conditional Probability

Conditional probability is defined as the probability that a second event will happen, given that the first event has already occurred. When two events are dependent, the conditional probability of A given B is

$$P(A \text{ given } B) = P(A|B) = \frac{P(A \text{ and } B)}{P(B)} = \frac{P(A \cap B)}{P(B)}.$$

Example 10: A pair of dice is rolled. What is the probability that the sum of two dice will be greater than 7, A, if the first die rolled is a 4, B?

Step 1: Find A and B.
A = total of two dice > 7
$B = 4$

Step 2: Find the number of outcomes for the sum of the dice. There are three outcomes, since the first die must be four and the sum of the dice must be greater than 7.
A and $B = A \cap B = (4,4), (4,5), (4,6)$

Step 3: Find $P(A \text{ and } B)$ and $P(B)$.

$$P(A \text{ and } B) = \frac{\text{\# of outcomes for } A \text{ and } B}{\text{total \# of outcomes}} = \frac{3}{36} = \frac{1}{12}$$

$$P(B) = \frac{\text{\# of outcomes for } B}{\text{total \# of outcomes}} = \frac{1}{6}$$

$$P(A|B) = \frac{P(A \text{ and } B)}{P(B)} = \frac{\frac{1}{12}}{\frac{1}{6}} = \frac{1}{2}$$

The probability of rolling the sum of two dice that is greater than 7, given that the first dice must be 4 is $\frac{1}{2}$.

Example 11: The probability that UGA and GT both win a football game in the same weekend is 37%. The probability that just UGA wins is 68%. What is the probability that GT will win given that UGA has already won?

$$P(A|B) = \frac{P(A \text{ and } B)}{P(B)} = \frac{37\%}{68\%} = 54\%$$

Find the conditional probability.

1. Kelly took two math tests. Her teacher said 30% of the class passed both tests, but 52% passed the first test. What is the probability that Kelly, who passed the first test, also passed the second test?

2. In California, 88% of teenagers have a cell phone and 76% have a cell phone and a MP3 player. What is the probability that a teenager has a MP3 player given that he or she also has a cell phone?

3. Tonya is selling 2 kinds of cookies, chocolate chip and sugar. After going to 100 houses, she determines that 41% of people bought chocolate chip cookies and 38% bought chocolate chip and sugar cookies. What percentage of people bought sugar cookies given that they already purchased chocolate chip cookies?

4. Lenny and Rebecca bought two scratch off lottery tickets. The probability that they both win is 6%. The probability that just one of them wins is 13%. What is the probability that Lenny will win given the fact that Rebecca already won?

5. The Braves have a double-header on Saturday. The probability that they will win both games is 31%. The probability that they will win just the first game is 71%. What is the probability they will win the second game given that they have already won the first game?

6. During a storm, the probability of the power and the cable going out is 31%. The probability that just the cable will go out is 54%. What is the probability that the power will go out given that the cable is already out?

7. The Nutcracker is being performed at the Fox Theater. The probability that the show will sell out on Friday and Saturday night is 61%. The probability that it will sell out Friday night is 63%. What is the probability that the show will sell out Saturday night if it sold out Friday night?

8. During a race, the probability that a racecar driver will have two flat tires is 13%. The probability that he will have one flat tire is 21%. Given that he has already had one flat tire, what is the probability that he will have another?

9. The probability of thunder and lightning during a summer storm is 19%. The probability of only having thunder is 27%. What is the probability that lightning will occur given that there is already thunder?

10. Matthew and Lexi decided to go out to dinner. The probability that they both order a hamburger is 7%. The probability that one of them orders a hamburger is 24%. What is the probability that Matthew will order a hamburger if Lexi has already ordered one?

11. A pair of dice is rolled. What is the probability that the sum of two dice will be greater than 9 if the first die rolled is a 6?

12. A pair of dice is rolled. What is the probability that the sum of two dice will be less than 5 if the second die rolled is a 2?

11.8   Predictions Using Probabilities

Once you know the probability of an event, you can use that probability to make predictions about future events.

Example 12:   Slugger Jones has a 0.475 batting average this season, meaning that he gets a hit 47.5% of his official at-bats (excluding walks, hit by pitch, etc.). In the double-header yesterday, he had 11 official at-bats. How many hits would you have expected him to collect?

Solution: Multiply the probability by the number of total opportunities. $0.475 * 11 = 5.225$ We would expect him to have gotten 5 hits in 11 at-bats.

Example 13:   Farmer Jacob owns so many sheep that he has 5 assistant shepherds. He wants to check on the composition of his flock, so he selects 40 sheep at random to see what they look like. Striped sheep are the most valuable. The results of his survey are below:

Type of sheep	White	Black	Striped	Speckled
Number of sheep	15	12	7	6

If one sheep is chosen at random, what is the probability that it is striped? If Farmer Jacob has 750 sheep total, how many of his sheep will probably be striped?

Step 1:   The probability of one sheep being striped is
$$\frac{\text{The number of striped sheep}}{\text{The total number of sheep}} = \frac{7}{40} = 17.5\%$$

Step 2:   Multiply the probability of one sheep being striped by the total number of sheep in the flock:
$0.175 \times 750 = 131.25$
He most likely has 131 striped sheep.

Example 14:   Mr. Gallop takes a poll of a randomly selected portion of the 10th grade class about their favorite baseball team. He makes a chart with the top 5 responses:

Top 5 Responses Among 10th Graders to the Question, "Who is Your Favorite Baseball Team?"

Favorite Team	Baltimore	Washington, DC	New York	Chicago	Boston
# of Responses	24	18	7	6	5

Mr. Gallop reads this chart and counts that there are 63 responses listed. He concludes that there is a $\frac{7}{63} = \frac{1}{9}$ probability that a randomly selected student would be a New York fan.
Therefore, he says, out of the 432 students in the 10th grade, $432 \times \frac{1}{9} = 48$ are likely to be New York fans. Why is his conclusion invalid?

Solution: The chart only gives the Top 5 responses about 10th graders to Mr. Gallop's question, not every response. Therefore, we do not know how many total students were polled, and we cannot calculate probabilities or make predictions.

Use the probabilities in the questions below to make predictions, if possible. If it is not possible to make predictions, explain why.

1. Mrs. Polley's oven burns about 6% of the cookies in a batch, on average. She makes 400 cookies for her daughter Lynn's wedding. How many cookies should she expect her oven to burn?

2. Christie Jo is on the Prom Planning Panel (PPP). She does a random survey of 30 female classmates to find out their plans for prom. The results are shown below.

Prom Plans	Have a Date	Hope to Have a Date	Attending without Date	Not Attending
# of Females	5	13	3	9

What is the probability that a female student selected at random is planning to attend? If there are 340 female students eligible to go to the prom, how many should Christie Jo expect to attend (based on this survey)?

3. Prisci surveyed a random sample of her classmates about their favorite type of ice cream. The results are shown below.

Favorite Types of Ice Cream at Grosami High School

Favorite Ice Cream*	Vanilla (incl. French Vanilla)	Chocolate	Strawberry	Chocolate Chip
# of Responses	24	8	5	3

* lists only ice cream flavors that received 3 or more responses

According to the survey results, what is the probability that a randomly selected student would prefer chocolate? If there are 80 students total in Prisci's class, how many of them would she predict prefer chocolate?

4. Van uses the SureFire cell phone company. They claim to have a dropped call rate of 5% in his area. If Van plans on making 60 phone calls this month, how many dropped calls should he expect?

5. Peggy takes a lot of photographs, and she prefers the quality of old-style film cameras. She buys rolls of film that take 60 pictures each. The table below shows the number of photos from each roll that Peggy rejects from putting in a photo album.

Peggy's Rejected Photos (rolls of 60)

Film Roll	1	2	3	4	5	6	7
Rejected Photos	3	8	2	0	4	5	5

In a year, Peggy uses 40 rolls, or 2400 photos. Based on these 7 rolls, how many photos would you expect her to reject? Round to the nearest whole number.

11.9   Expected Value

The expected value of a variable is the same as the average or mean of the variable. If we have a variable, X, the expected value is $E(X) = \sum_{i=1}^{n} X_i P_i$ where P_i is the probability of the value of X_i.

Example 15:   Suppose that the following game is played. A man rolls a fair die. If he rolls 1, 3, or 5, he loses $3, if he rolls 4 or 6, he loses $2, and if he rolls 2, he wins $12. What gain (loss) should he expect on average? (What is his expected value?)

Step 1:   We can define a random variable X to be the amount of money associated with each outcome.

Therefore, the possible values of X are $-\$3$ (if 1, 3, or 5 is rolled), $-\$2$ (if 4 or 6 is rolled), and $\$12$ (if 2 is rolled).

Note: $3 and $2 are negative because he loses that money, not gains.

Step 2:   Find the probabilities of each of the outcomes, $P(-\$3)$, $P(-\$2)$, and $P(\$12)$. To do this make a table.

Possible values of X	Probability $P(X)$
$-\$3$	P (rolling a 1, 3, or 5) $= \frac{3}{6}$
$-\$2$	P (rolling a 4 or 6) $= \frac{2}{6}$
$\$12$	P (rolling a 2) $= \frac{1}{6}$

Step 3:   Now, we multiply the probability for each outcome by the amount of money associated with that specific outcome.

Expected Value of Rolling a 1, 3, or, 5: $-\$3 \times \frac{3}{6} = -\$1.5$

Expected Value of Rolling a 4 or 6: $-\$2 \times \frac{2}{6} = -\$0.67$

Expected Value of Rolling a 2: $\$12 \times \frac{1}{6} = \$2$

Step 4:   To find the expected value of the entire game, add up the expected value for each outcome.

$$E(X) = (-\$3)\left(\tfrac{3}{6}\right) + (-\$2)\left(\tfrac{2}{6}\right) + (\$12)\left(\tfrac{1}{6}\right)$$

$$E(X) = -\$1.5 - \$\tfrac{2}{3} + \$2 = -\$\tfrac{1}{6}$$

The expected value is $-\$\frac{1}{6}$.

This means that the man playing the game is expected to lose an average of 16.7 cents $\left(\frac{1}{6} \text{ of } \$1.00\right)$ each game he plays.

Find the expected value.

1. Ken tells you to throw a die 100 times and he will give you whatever you roll in dollars (if you roll a 1, you get $1 etc.). Before you play, you want to estimate how much money Ken will end up giving you.

2. Mark and Rodney are playing a card game. Every time Mark draws a Jack or Queen, Rodney gives him $4. When Mark draws a King or Ace, Rodney gives him $5. In turn, Mark gives Rodney $X for every other card drawn. In order for the game to be fair (the expected value is zero), how much should Mark pay Rodney for the cards he draws? (Hint: Find the value of X.)

3. On the first day of class, Kara's teacher gave the class the following chart to represent the test number and the probable percent of people that will make an A on each test.

Test	1	2	3	4	5	6	7	8	9	10
P(A)	0.36	0.32	0.28	0.25	0.21	0.19	0.16	0.14	0.09	0.03

How many people out of a class of 40 will make an A? Round to the nearest whole number.

4. Suppose that the following game is played. A man rolls a fair die. If he rolls 2 or 5, he loses $5, if he rolls 1 or 3, he loses $7, and if he rolls 4 or 6, he wins $15. What gain (loss) should he expect on average?

5. Suppose that the following game is played. A woman rolls a fair die. If she rolls 1, she loses $15, if she rolls 2 or 3, she loses $10, and if she rolls 4, 5, or 6, she wins $20. What gain (loss) should she expect on average?

6. Suppose that the following game is played. A man rolls a fair die. If he rolls 1 or 5, he loses $3, if he rolls 2, 3, or 6, he loses $5, and if he rolls 4, he wins $11. What gain (loss) should he expect on average?

7. Suppose that the following game is played. A woman rolls a fair die. If she rolls 1 or 6, she loses $1, if she rolls 2 or 5, she loses $2, and if she rolls 3 or 4, she wins $9. What gain (loss) should she expect on average?

8. Suppose that the following game is played. A man rolls a fair die. If he rolls 1, he loses $6, if he rolls 2, 3, or 4, he loses $3, if he rolls 5, he loses $5, and if he rolls 6, he wins $18. What gain (loss) should he expect on average?

Chapter 11 Review

1. There are 50 students in the school orchestra in the following sections:

string section	woodwind	percussion	brass
25	15	5	5

   One student will be chosen at random to present the orchestra director with an award. What is the probability the student will be from the woodwind section?

2. Fluffy's cat treat box contains 6 chicken-flavored treats, 5 beef-flavored treats, and 7 fish-flavored treats. If Fluffy's owner reaches in the box without looking, and chooses one treat, what is the probability that Fluffy will get a chicken-flavored treat?

3. The spinner in figure A stopped on the number 5 on the first spin. What is the probability that it will not stop on 5 on the second spin?

   Fig. A   Fig. B

4. Sherri turns the spinner in figure B above 3 times. What is the probability that the pointer always lands on a shaded number?

5. Three cakes are sliced into 20 pieces each. Each cake contains 1 gold ring. What is the probability that one person who eats one piece of cake from each of the 3 cakes will find 3 gold rings?

6. Brianna tosses a coin 4 times. What is the probability she gets all tails?

7. Tempest has a bag with 4 red marbles, 3 blue marbles, and 2 yellow marbles. Does adding 4 purple marbles increase or decrease her chances that the first marble she draws at random will be red?

8. Simone has lived in Silver Spring for 250 days, and in that time her power has been out 3 days.
   (A) If the power outages happen at random, what is the probability that the power will be out tomorrow?
   (B) If the probability remains the same, how many days will she be without power for the next 10 years (3,652 days)? Round to the nearest whole number.

9. There are 20 balls in a box. 12 are green, 8 are black. If two balls are chosen without replacement, what is the probability that both are green balls?

10. Suppose that the following game is played. A man rolls a fair die. If he rolls 2, 4, or 6, he loses $2, if he rolls 1 or 3, he loses $5, and if he rolls 5, he wins $9. What gain (loss) should he expect on average? (What is his expected value?)

11. During football season, the probability of students missing school is 16%. The probability that they will miss school and go to the football game is 9%. What is the probability a student will go to the game if they miss school on Friday?

12. At Texas State University, the probability that a student is taking Spanish is 72%. The probability that a student is taking Spanish and Self Defense is 28%. What is the probability that a student is taking Self Defense given that they are already registered to take Spanish?

13. Branden has a box of 50 cookies, 4 of which are peanut butter. On the way home he randomly picks two cookies to eat as his snack. What is the probability he chose

    (A) 2 peanut butter cookies (no replacement)?
    (B) 2 peanut butter cookies if he puts the first one back?
    (C) 1 peanut butter cookie and 1 other cookie (no replacement)?

14. At the county fair, there are 5 unlabeled cakes. 3 are on table A. 2 of those 3 were made by Ann and the third was made by Betty. Two cakes (one of Ann's and one of Betty's) are on table B. The judge at the fair moves one cake from table A to table B. Jack then randomly chooses a cake from table B. What is the probability that he chose a cake made by Betty?

Read the following, and answer questions 15–19.

There are 9 slips of paper in a hat, each with a number from 1 to 9. The numbers correspond to a group of students who must answer a question when the number for their group is drawn. Each time a number is drawn, the number is put back in the hat.

15. What is the probability that the number 6 will be drawn twice in a row?

16. What is the probability that the first 5 numbers drawn will be odd numbers?

17. What is the probability that the second, third, and fourth numbers drawn will be even numbers?

18. What is the probability that the first five times a number is drawn it will be the number 5?

19. What is the probability that the first five numbers drawn will be 1, 2, 3, 4, 5 in that order?

Solve the following word problems. For questions 20–22, write whether the problem is "dependent" or "independent".

20. Felix Perez reaches into a 10-piece puzzle and pulls out one piece at random. This piece has two places where it could connect to other pieces. What is the probability that he will select another piece which fits the first one if he selects the next piece at random?

21. Barbara Stein is desperate for a piece of chocolate candy. She reaches into a bag which contains 8 peppermint, 5 butterscotch, 7 toffee, 3 mint, and 6 chocolate pieces and pulls out a toffee piece. Disappointed, she throws it back into the bag and then reaches back in and pulls out one piece of candy. What is the probability that Barbara pulls out a chocolate piece on the second try?

22. Christen Solis goes to a pet shop and immediately decides to purchase a guppy she saw swimming in an aquarium. She reaches into the tank containing 5 goldfish, 6 guppies, 4 miniature catfish, and 3 minnows and accidently pulls up a goldfish. Breathing a sigh, Christen places the goldfish back in the water. The fish are swimming so fast, it is impossible to tell what fish Christen would catch. What is the probability that Christen will catch a guppy on her second try?

Chapter 11 Test

1. There are 10 boys and 12 girls in a class. If one student is selected at random from the class, what is the probability it is a girl?

(A) $\frac{1}{2}$

(B) $\frac{1}{22}$

(C) $\frac{6}{11}$

(D) $\frac{6}{5}$

2. David just got a new credit card in the mail. What is the probability the second digit of the credit card number is a 3?

(A) $\frac{1}{3}$

(B) $\frac{1}{10}$

(C) $\frac{2}{3}$

(D) $\frac{1}{5}$

3. Brenda has 18 fish in an aquarium. The fish are the following colors: 5 orange, 7 blue, 2 black, and 4 green. Brenda also has a trouble-making cat that has grabbed a fish. What is the probability the cat grabbed a green fish if all the fish are equally capable of avoiding the cat?

(A) $\frac{2}{9}$

(B) $\frac{1}{18}$

(C) $\frac{2}{7}$

(D) $\frac{1}{4}$

4. In problem number 3, what is the probability the cat will **not** grab an orange fish?

(A) $\frac{1}{3}$

(B) $\frac{13}{18}$

(C) $\frac{3}{4}$

(D) $\frac{13}{5}$

5. You have a cube with each face numbered 1, 2, 3, 4, 5, or 6. What is the probability if you roll the cube 4 times, you will get the number 5 each time?

(A) $\frac{4}{1296}$

(B) $\frac{4}{5}$

(C) $\frac{1}{256}$

(D) $\frac{1}{1296}$

6. In a game using two numbered cubes, what is the probability of **not** rolling the same number on both cubes three times in a row?

(A) $\frac{125}{256}$

(B) $\frac{125}{216}$

(C) $\frac{27}{64}$

(D) $\frac{16}{27}$

7. Carrie bought a large basket of 60 apples. When she got them home, she found 4 of the apples were rotten. If she goes back and buys 200 more apples, about how many rotten apples should she expect?

(A) 8
(B) 13
(C) 20
(D) 24

8. Katie spun a spinner 15 times and recorded her results in a table below. The spinner was divided into 6 sections numbered 1–6. The results of the spins are shown below.

```
1  3  6  5  1
6  2  4  3  4
2  5  1  4  5
```

Based on the results, how many times would 4 be expected to appear in 45 spins?

(A) 9
(B) 12
(C) 15
(D) 21

9. Ben has a bag of 40 red triangles and 40 blue triangles. Without replacement, what is the probability of Ben choosing 2 red triangles?

(A) 0.20
(B) 0.25
(C) 0.30
(D) 0.35

10. Kyle has 4 different kinds of Halloween candy in a bag. He has 5 chocolate rolls, 6 lollipops, 4 chocolate bars, and 7 peanut butter candies. What kind of candy must Kyle have picked out of the bag if the probability of picking that kind is $\frac{2}{11}$?

(A) chocolate roll
(B) lollipop
(C) chocolate bar
(D) peanut butter

11. There are 7 sections of equal size on a spinner. One is labeled purple. Two are labeled green. Three are labeled orange. One section is unlabeled. What color must the unlabeled section be if the probability of spinning that color is $\frac{4}{7}$?

(A) purple
(B) green
(C) orange
(D) brown

12. The table below shows the actual sum of the rolling of two cubes numbered 1 through 6. The two cubes were rolled 100 times.

Sum	Frequency
2	5
3	5
4	9
5	10
6	15
7	14
8	15
9	11
10	9
11	4
12	3

Using the information in the table, predict how many times a score of "7" would occur in 150 tries.

(A) 21
(B) 28
(C) 35
(D) 43

13. In Florida, 94% of people have a TV. 71% have a TV and a VCR. What percent of people have a VCR if they already had a TV?

(A) 66%
(B) 70%
(C) 83%
(D) 76%

14. If there are ten blocks in a bucket, 7 are red and 3 are yellow, what are the odds of choosing 2 red blocks (without replacement)?

(A) $\dfrac{8}{15}$

(B) $\dfrac{7}{30}$

(C) $\dfrac{6}{15}$

(D) $\dfrac{7}{15}$

15. If there are 100 blocks in a bucket, 40 are black, 35 are white, 20 are silver, and 5 are clear, what are the odds of choosing 6 clear blocks?

(A) 0

(B) $\dfrac{1}{2}$

(C) $\dfrac{1}{20}$

(D) $\dfrac{1}{10}$

16. Karen has two tests this week. The chance that she'll pass both is 62%. The chance she'll pass one is 74%. What's the chance she'll pass one if she already passes the other?

(A) 80%

(B) 87%

(C) 89%

(D) 84%

17. In high school, 23% of boys play football. 16% play football and basketball. What percent of boys play basketball if they are already playing football?

(A) 50%

(B) 70%

(C) 80%

(D) 30%

18. Sarah has a bag of 100 marbles. 40 are blue, 40 are green, and 20 are purple. What is the probability of Sarah picking a purple marble on the first try?

(A) $\dfrac{1}{5}$

(B) $\dfrac{1}{4}$

(C) $\dfrac{1}{100}$

(D) $\dfrac{8}{25}$

Chapter 12
Statistics

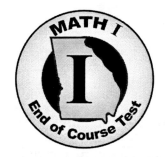

This chapter covers the following Georgia Performance Standards:

MM1D	Data Analysis and Probability	MM1D3a
		MM1D3b
		MM1D3c
		MM1D4

12.1   Mean

In statistics, the arithmetic mean is the same as the average. To find the arithmetic mean of a list of numbers, first add together all of the numbers in the list, and then divide by the number of items in the list.

Example 1:   Find the mean of 38, 72, 110, 548.

  Step 1:   First add: $38 + 72 + 110 + 548 = 768$

  Step 2:   There are 4 numbers in the list so divide the total by 4. $768 \div 4 = 192$
              The mean is 192.

Practice finding the mean (average). Round to the nearest tenth if necessary.

1. Dinners served:
    489 561 522 450

2. Prices paid for shirts:
    $4.89 $9.97 $5.90 $8.64

3. Piglets born:
    23 19 15 21 22

4. Student absences:
    6 5 13 8 9 12 7

5. Paychecks:
    $89.56 $99.99 $56.54

6. Choir attendance:
    56 45 97 66 70

7. Long distance calls:
    33 14 24 21 19

8. Train boxcars:
    56 55 48 61 51

9. Cookies eaten:
    5 6 8 9 2 4 3

Find the mean (average) of the following word problems.

10. Val's science grades are 95, 87, 65, 94, 78, and 97. What is her average?

11. Ann runs a business from her home. The number of orders for the last 7 business days are 17, 24, 13, 8, 11, 15, and 9. What is the average number of orders per day?

12. Melissa tracks the number of phone calls she has per day: 8, 2, 5, 4, 7, 3, 6, 1. What is the average number of calls she receives?

13. The Cheese Shop tracks the number of lunches they serve this week: 42, 55, 36, 41, 38, 33, and 46. What is the average number of lunches served?

14. Leah drives 364 miles in 7 hours. What is her average miles per hour?

15. Tim saves $680 in 8 months. How much does his savings average each month?

16. Ken makes 117 passes in 13 games. How many passes does he average per game?

12.2   Finding Data Missing From the Mean

Example 2:   Mara knew she had an 88 average in her biology class, but she lost one of her papers. The three papers she could find had scores of 98%, 84%, and 90%. What was the score on her fourth paper?

Step 1:   Calculate the total score on four papers with an 88% average. $.88 \times 4 = 3.52$

Step 2:   Add together the scores from the three papers you have. $.98 + .84 + .9 = 2.72$

Step 3:   Subtract the scores you know from the total score. $3.52 - 2.72 = .80$. She had 80% on her fourth paper.

Find the data missing from the following problems.

1. Gabriel earns 87% on his first geography test. He wants to keep a 92% average. What does he need to get on his next test to bring his average up?

2. Rian earned $68.00 on Monday. How much money must she earn on Tuesday to have an average of $80 earned for the two days?

3. Haley, Chuck, Dana, and Chris enter a contest to see who could bake the most chocolate chip cookies in an hour. They bake an average of 75 cookies. Haley bakes 55, Chuck bakes 70, and Dana bakes 90. How many does Chris bake?

12.3   Median

In a list of numbers ordered from lowest to highest, the **median** is the middle number. To find the **median**, first arrange the numbers in numerical order. If there is an odd number of items in the list, the **median** is the middle number. If there is an even number of items in the list, the **median** is the **average of the two middle numbers.**

Example 3:   Find the median of 42, 35, 45, 37, and 41.

Step 1:   Arrange the numbers in numerical order: 35 37 41 42 45

Step 2:   Find the middle number. The median is 41.

Example 4:   Find the median of 14, 53, 42, 6, 14, and 46.

Step 1:   Arrange the numbers in numerical order: 6 14 14 42 46 53.

Step 2:   Find the average of the two middle numbers. $(14 + 42) \div 2 = 28$. The median is 28.

Circle the median in each list of numbers.

1. 35, 55, 40, 30, and 45

2. 7, 2, 3, 6, 5, 1, and 8

3. 65, 42, 60, 46, and 90

4. 15, 16, 19, 25, 20

5. 75, 98, 87, 65, 82, 88, 100

6. 33, 42, 50, 22, and 19

7. 10, 8, 21, 14, 9, and 12

8. 43, 36, 20, and 40

9. 5, 24, 9, 18, 12, and 3

10. 48, 13, 54, 82, 90, and 7

11. 23, 21, 36, and 27

12. 9, 4, 3, 1, 6, 2, 10, and 12

12.4   Quartiles and Extremes

In statistics, large sets of data are separated into four equal parts. These parts are called **quartiles**. The **median** separates the data into two halves. Then, the median of the upper half is the **upper quartile**, and the median of the lower half is the **lower quartile**. The distance between the upper quartile and the lower quartile is the **interquartile range**. The interquartile range is sometimes used in the place of range, especially when there are outliers in the data set. Interquartile range is also another type of variability of the data.

The **extremes** are the highest and lowest values in a set of data. The lowest value is called the **lower extreme**, and the highest value is called the **upper extreme**.

Example 5:   The following set of data shows the high temperatures (in degrees Fahrenheit) in cities across the United States on a particular autumn day. Find the median, the upper quartile, the lower quartile, the upper extreme, and the lower extreme of the data.

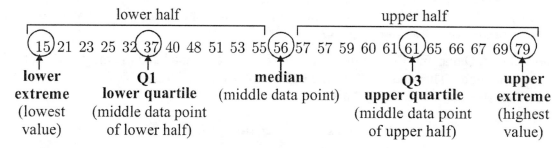

Example 6:   The following set of data shows the fastest race car qualifying speeds in miles per hour. Find the median, the upper quartile, the lower quartile, the upper extreme, and the lower extreme of the data.

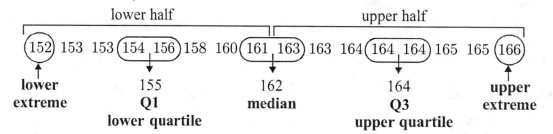

Note: When you have an even number of data points, the median is the average of the two middle points. The lower middle number is then included in the lower half of the data, and the upper middle number is included in the upper half.

Find the median, the upper quartile, the lower quartile, the upper extreme, the lower extreme, and interquartile range of each set of data given below.

1. 0 0 1 1 1 2 2 3 3 4 5

2. 15 16 18 20 22 22 23

3. 62 75 77 80 81 85 87 91 94

4. 74 74 76 76 77 78

5. 3 3 3 5 5 6 6 7 7 7 8 8

6. 190 191 192 192 194 195 196

7. 6 7 9 9 10 10 11 13 15

8. 21 22 24 25 27 28 32 35

12.5   Comparing Statistics

The best way to compare statistics is to use summary statistics. To compare, you could use the mean, median, mode, range, interquartile range, or quartiles.

Example 7:   Compare Zack and Cody's test scores using the mean, median, mode, range, interquartile range, and quartiles.

Zack	81	72	91	88	71	73	82	100	81	86
Cody	86	80	86	84	72	80	90	85	89	92

Compare the **mean** of Zack and Cody's test scores. Whose average is higher?

Zack: $\dfrac{81 + 72 + 91 + 88 + 71 + 73 + 82 + 100 + 81 + 86}{10} = \dfrac{825}{10} = 82.5$

Cody: $\dfrac{86 + 80 + 86 + 84 + 72 + 80 + 90 + 85 + 89 + 92}{10} = \dfrac{844}{10} = 84.4$

Cody's test average is higher.

Note: The mean (average) is the only value that is sensitive to extreme values. The process of comparing averages is used for samples that are very large. In cases where the sample is very large, we call the estimate of that data a **parameter**. In cases where we take the time to measure the entire population, the parameter becomes known as the **population parameter**.

Now compare their test scores using the **median**.
Zack's median $= 81.5$ Cody's median $= 85.5$
Cody's average is still higher!

Now compare their test scores using the **mode**.
Zack's mode $= 81$, because it appears twice in his test scores
Cody does not have a MODE because he has two sets of numbers that both appear twice $(80, 86)$

Now compare their test scores using the **range**.
Zack's range $= 100 - 71 = 29$ Cody's range $= 92 - 71 = 20$
Cody's range is smaller meaning his grades are a little more localized around one point, or grade.

Now compare their test scores using the **interquartile range**. (Remember: The interquartile range (IQR) is the measure of variance between the 3rd and 1st quartiles.)

Zack's IQR $= Q3 - Q1 = 88 - 73 = 15$. There is a 15 point spread (variability) in grades.
Cody's IQR $= Q3 - Q1 = 89 - 80 = 9$. There is a 9 point spread (variability) in grades.

It is important to know that the mean, median, and mode are measures of the **center**. The range and interquartile range are measures of the **spread (variability)**.

Statistics can also be used to compare a small group of data to a large group of data.

Example 8:   Stephen is a student at North Cobb High School. Look at the following histograms. The first histogram displays the amount of money earned each week by each student (rounded to the nearest ten) in Stephen's English class. The second histogram displays the amount of money earned each week by each of the students (rounded to the nearest ten) in the ninth-grade class at his school.

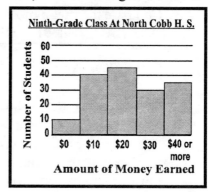

    a. Compare the medians of the data in the histograms.
    b. Compare the lower quartiles of the data in the histograms.
    c. Compare the modes of the data in the histograms.

Solution:

  a. To find the median of Stephen's English class, first find the total number of students in his class. By adding up the number of students in each bar $(4 + 6 + 5 + 3 + 5)$, we find there are 23 students in Stephen's English class. The median is the middle number in the data set that is written from least to greatest. The middle number is Stephen's English class is 12, which is found within the third bar ($20).

    The total number of students in ninth-grade at North Cobb High School is found by adding up the number of students in each bar $(10 + 40 + 45 + 40 + 35)$. There are 170 students in ninth-grade at North Cobb High School. The median is between the 85th and 86th student, which is found within the third bar ($20).

    The median of the amount of money earned by each of the students in the ninth-grade class at North Cobb High School is the same as the median of the amount of money earned by each of the students in Stephen's English class.

  b. The lower quartile is the median of the first half of data.
    The first half of data in Stephen's English class includes the first 12 students, so the median of this is found within the second bar ($10).
    The first half of data in ninth-grade at North Cobb High School includes the first 85 students, so the median of this is found with the second bar ($10).
    The lower quartile of the amount of money earned by each of the students in the ninth-grade class at North Cobb High School is the same as the lower quartile of the amount of money earned by each of the students in Stephen's english class.

c. The mode is easy to find. The mode is the value that is list the most.
    In Stephen's English class, the mode is $10.
    In the ninth-grade class at North Cobb High School, the mode is $20.
    The mode of the amount of money earned by each of the students in the ninth-grade class
    at North Cobb High School is greater than the mode of the amount of money earned by
    each of the students in Stephen's English class.

Use the chart below for problems 1–5.

| Sample A | 60 | 71 | 73 | 69 | 80 | 82 |
| Sample B | 71 | 74 | 73 | 79 | 81 | 80 |

1. Find the means.

2. Find the modes.

3. Find the medians.

4. Find the ranges.

5. Find the interquartile range.

Use the bar graph below for problems 6–10.

6. Find the means.

7. Find the modes.

8. Find the medians.

9. Find the ranges.

10. Find the interquartile range.

12.6   Simple Random Sampling

An important sampling technique is called **simple random sampling**. Simple random sampling is often used because it provides an excellent **representative sample** of the overall population. A representative sample is a sample that reflects (or represents) the characteristics of the whole population. There are two elements that make a sample a "simple random sample."

The first is that the sample is **unbiased**. In an unbiased sample, each member of the population is equally likely to be chosen, so the sample is representative of the population.

The second element of simple random sampling is that the observations are **independent.** In mathematics, the outcome of an event may or may not influence the outcome of a second event. If the outcome of one event does not influence the outcome of the second event these events are independent. However, if one event has an influence on the second event, the events are **dependent**.

Example 9:   Jeremy wanted to take a simple random sample of the 9th grade's movie preferences. To do so, he gave everyone in the 9th grade a ticket with a number on it, and randomly chose 15 numbers out of a hat. Then Jeremy had each of them fill out a slip of paper with their favorite movie, and he had each of them get one friend to fill out a slip with their favorite movie. Was this a simple random sample?

Solution: No, it is not a simple random sample, because the observations are not independent. The choice of the 15 friends is dependent on the 15 students who were drawn out of the hat. Jeremy should simply draw all 30 names out of the hat.

Determine whether the following samples are simple random samples, and if not, whether they are biased or dependent.

1. Henry Higgins K–8 School conducts a survey to determine children's attitudes about their parents. They get out a list of students in alphabetical order and choose every 7th name on the list to survey.

2. Million Milk tests how long it takes their milk to spoil. They use a random number generator to select 100 one-gallon bottles from among all of their products to test.

3. Reverend Nance conducts a survey on American attitudes about religion. He assigns everyone who comes to his church every Sunday a number. The reverend uses a random number generator to select 150 of them to take the survey.

4. A social scientist conducts a study on the length of time that Baltimore County newlyweds of last year dated before they married. She has the phone numbers of everyone who obtained a marriage license from Baltimore County courthouse last year, and she programs her computer to select at random 100 couples to call and survey.

Answer the following question in your own words.

5. The Cuckoo Cola Company wants to gain a sense of their workers' satisfaction, so they decide to survey some of their $17,000$ U.S. employees. Based on simple random sample principles, explain at least two of the key elements that should be part of the administration's design.

12.7   Sample Size

A final important consideration for sampling is sample size. The general rule is, the bigger the sample, the more accurate the results (as long as the sample is a representative sample).

A good rule of thumb is that for populations of 50 or smaller, you want a simple random sample size of at least $\frac{3}{4}$ or 75% of the population.

For populations larger than 50 but smaller than 100, you want a simple random sample size of at least $\frac{1}{2}$ or 50%.

For populations larger than 100, a simple random sample size of 50 or 60 is the least you want.

These rules of thumb are not exact but will help you if a question asks if the "sample size is too small." Some questions will ask you to design an experiment, and you should always include that "sample size should be large enough" and sometimes specify a sample size that is large enough.

Example 10:   Eustace came to his teacher, saying, "Mr. Wehrenberg, I just did a simple random sample and found that 50% of our class hates our school and wants the principal fired!" Mr. Wehrenberg asked Eustace how many people he had surveyed. Eustace responded, "Two."
There are 32 people in the class. How many should Eustace survey to have a sufficient sample size, according to the rule of thumb?
There are 700 people in the school. How many of those would Eustace have to survey to have a sufficient sample size?

Step 1:   The class has 32 students, which is less than 50, so according to the rule of thumb, he should want 75% which would be 24 students.

Step 2:   For the whole school, he should survey at least 50 or 60 students, although the more he could survey, the more accurate his results would be.

Answer the following questions.

1. True or false. Increasing the size of a simple random sample always increases its accuracy.

2. True or false. For a population of 78, a sample size of 20 would be as much as you would want for accuracy

3. True or false. For a population of 43, you should be content with a sample size of 35.

4. True or false. For a population of 50,000, a sample size of 100 should be fairly accurate.

12.8   Representative Sampling

Another important sampling technique is called **representative sampling**. Often, taking a simple random sample is not possible, either because there is no way to obtain a totally unbiased sample or a totally independent sample. The goal is still to obtain a sample as representative of the population as possible. A key element of representative sampling is **variety.**

Example 11:   Emma works for a company that produces stuffed animals at 5 factories across the Midwest. She wants to test the quality of the stuffed animals just as they come off the factory line (to eliminate the possibility of damage during shipping). Which of the following experiment designs would give the most representative sample?

(A) Go to the nearest factory and choose 100 stuffed animals at random to test.

(B) Go to each of the 5 factories and take the first 20 stuffed animals that come off the factory line.

(C) Go to each of the 5 factories and choose 20 stuffed animals at random to test.

Solution: Option (A) is a random sample that gives an excellent representation of the population of stuffed animals at that particular factory. However, Emma wants a sample that represents the population of stuffed animals at all 5 factories.

Option (C), on the other hand, is a sample that is drawn from all 5 factories. It has greater variety, so it will be more representative.

Option (B) also draws from all 5 factories, but by choosing consecutive stuffed animals, it is not an independent sample. For instance, the machines and workers may produce better stuffed animals early in the day or the week.

Option (C) is the most representative sample.

Answer the following question about representative sampling and explain your answer.

1. ReallyGoodJobs.com conducts a survey to determine the attitudes of college seniors about entering the workforce. Which of these sampling techniques would give them the most representative samples?

   (A) Send surveyors to 15 randomly selected college campuses across the country and ask every person who attends the school's women's soccer game to fill out a survey.

   (B) Send surveyors to interview all 600 seniors at NW Maryland State University.

   (C) Choose 15 randomly selected college campuses across the country and use a computer to randomly select 40 seniors from each school out of the school's student directory.

   (D) Put the survey on their website and wait until 600 college seniors respond to it.

12.9   Mean Absolute Deviation

The **mean absolute deviation** is used as a way to determine the variability of data. **Variability** (or variance) is the measure of the differences in things (scores, results, other variables, etc.) with the mean as a reference. To find the mean absolute deviation, you must compute the average of the sum of the absolute values of the deviations of the data. The deviation of element in a data set is the difference between the element and the mean of the data set. The formula for average absolute deviation for a set $\{x_1, x_2, ..., x_n\}$ is

$$\frac{1}{n} \sum_{i=1}^{n} |x_i - \overline{x}|,$$

where x_i are the elements in the set $\{x_1, x_2, ..., x_n\}$ and $\overline{x}$ is the mean of the data set.

Example 12:   Find the mean absolute deviation of the data set $\{10, 10, 40, 60, 80\}$.

Step 1:   First find the mean $\overline{x}$ of the data set.
$$\overline{x} = \frac{10 + 10 + 40 + 60 + 80}{5} = \frac{200}{5} = 40$$

Step 2:   Now, plug the values into the equation $\frac{1}{n} \sum_{i=1}^{n} |x_i - \overline{x}|$.

$$\frac{\sum_{i=1}^{n} |x_i - \overline{x}|}{n} = \frac{|10 - 40| + |10 - 40| + |40 - 40| + |60 - 40| + |80 - 40|}{5}$$

$$= \frac{|-30| + |-30| + |0| + |20| + |40|}{5} = \frac{30 + 30 + 0 + 20 + 40}{5} = \frac{120}{5} = 24$$

The mean absolute deviation of the data set is 24.
This means that the average distance between the elements and the mean of the data set is 24. (Some elements are closer to the mean and some are further away, but the average distance is 24.)

Given a data, find the mean absolute deviation.

1. $\{1, 2, 3, 4, 5\}$

2. $\{2, 6, 8, 8\}$

3. $\{0, 1, 1, 1, 4, 5\}$

4. $\{7, 14, 15\}$

5. $\{0, 2, 4, 4, 5\}$

6. $\{0, 10, 11\}$

7. $\{2, 2, 3, 5, 8\}$

8. $\{19, 20, 21\}$

9. $\{2, 8, 13, 17\}$

10. $\{2, 3, 5, 10, 15\}$

Chapter 12 Review

Find the mean, median, mode, and range for each of the following sets of data. Fill in the table below.

❶ Miles Run by Track Team Members

Jeff	24
Eric	20
Craig	19
Simon	20
Elijah	25
Rich	19
Marcus	20

❷ 1992 SUMMER OLYMPIC GAMES
Gold Medals Won

Unified Team	45	Hungary	11
United States	37	South Korea	12
Germany	33	France	8
China	16	Australia	7
Cuba	14	Japan	3
Spain	13		

❸ Hardware Store Payroll June Week 2

Erica	$280
Dane	$206
Sam	$240
Nancy	$404
Elsie	$210
Gail	$305
David	$280

Data Set Number	Mean	Median	Mode	Range
❶				
❷				
❸				

4. Jenica bowls three games and scores an average of 116 points per game. She scores 105 on her first game and 128 on her second game. What does she score on her third game?

5. Concession stand sales for each game in season are $320, $540, $230, $450, $280, and $580. What is the mean sales per game?

6. Cendrick D'Amitrano works Friday and Saturday delivering pizza. He delivers 8 pizzas on Friday. How many pizzas must he deliver on Saturday to average 11 pizzas per day?

7. Long cooks three Vietnamese dinners that weigh a total of 40 ounces. What is the average weight for each dinner?

8. The Swamp Foxes scored an average of 7 points per soccer game. They scored 9 points in the first game, 4 points in the second game, and 5 points in the third game. What was their score for their fourth game?

9. Shondra is 66 inches tall, and DeWayne is 72 inches tall. How tall is Michael if the average height of these three students is 67 inches?

Nine cooks are asked, "If you use a thermometer, what is the actual temperature inside your oven when it is set at 350°F?" The responses are in the chart below.

Temperature (°F)	104	347	348	349	350	351	352
Number of Cooks	1	1	1	2	1	2	1

10. Find the mean of the data above.

11. Find the median of the data above.

12. Captain Kendrick of the Starship Supernova wants to know how many of his crew of 700 are afraid of encountering intelligent aliens. Which of the following would give him a simple random sample?

(A) Ask all of his senior officers and friends.

(B) Tell the ship's computer to assign a number between 1–700 to every crewperson and use the computer to generate random numbers to select 70 of the crew for the survey.

(C) Get out the alphabetical crew listing and interview every 10th crew member.

(D) Choose 10 crew members randomly from each of the 7 halls in which his crew lives.

13. True or false? Increasing the sample size of a biased sample corrects for the bias.

14. A tree farmer has thousands of trees and cannot check the health of every one regularly. He wants to design a test to check the health of some of his trees in a way that they represent the whole population of trees. What are three considerations that should go into his test design?

Use the following to answer questions 15–20.

Over the past 2 years, Coach Strive has kept a record of how many points his basketball team, the Bearcats, has scored in 20 games:

29 32 35 36 38 39 40 40 41 42 43 44 44 45 47 49 50 52 53 62

15. What is the median?

16. What is the upper quartile?

17. What is the lower quartile?

18. What is the upper extreme?

19. What is the lower extreme?

20. What is the interquartile range?

Use the figure to help answer questions 21–27.

Jessica	92	85	97	87	88	86	91	93	89	89
Michael	91	89	83	95	80	91	81	96	93	93

21. What is the lowest score Jessica can receive on her next test in order to have a 90 test average?

22. What score does Michael need to make to get an A average?

23. Find the means.

24. Find the interquartile range for each person.

25. According to the different quartiles, whose Q1 is higher?

26. Find the ranges.

27. According to the results of the previous questions, who performed better throughout the year?

28. Find the mean absolute deviation of $\{14, 32\}$.

29. Find the mean absolute deviation of $\{2, 3, 4, 7\}$.

Chapter 12 Test

1. What is the mean of 36, 54, 66, 45, 36, 36, and 63?

   (A) 36
   (B) 45
   (C) 48
   (D) 63

2. What is the median of the following set of data?

   33, 31, 35, 24, 38, 30

   (A) 32
   (B) 31
   (C) 30
   (D) 29

3. Concession stand sales for the first 6 games of the season averaged $400.00. If the total sales of the first 5 games were $320, $540, $230, $450, and $280, what were the total sales for the sixth game?

   (A) $230
   (B) $350
   (C) $364
   (D) $580

4. What is the mean of 12, 23, 8, 26, 37, 11, and 9?

   (A) 12
   (B) 29
   (C) 18
   (D) 19

5. Which of the following sets of numbers has a median of 42?

   (A) {60, 42, 37, 22, 19}
   (B) {16, 28, 42, 48}
   (C) {42, 64, 20}
   (D) {12, 42, 40, 50}

6. The scores on the math quiz were 94, 73, 87, 81, 82, 62, 55, 60. What is the upper quartile value of these scores?

   (A) 77
   (B) 80
   (C) 84.1
   (D) 84.5

7. The cafeteria worker surveyed students in the lunch line about what was their favorite kind of pizza? The workers decide to survey the first 150 students that come into the cafeteria on Monday morning. Which of these **best** describes why this type of sampling may give biased results?

   (A) Students are randomly selected.
   (B) The sample size is too small.
   (C) Not all students have the same chance to complete the survey.
   (D) Every other student entering the cafeteria should be surveyed.

8. There are 72 students in the school orchestra this school year. They have been invited to do a concert for the President of the United States in Washington, which will be televised nation wide. The catch is the concert will be on Homecoming weekend. The music director wants to survey the orchestra about whether or not to accept the invitation. How many orchestra members should he survey to get a representative sample?

   (A) 12
   (B) 24
   (C) 36
   (D) 42

Use the figure for questions 9–10.

Bob	7	4	6	8	7	5	3
Ann	6	5	6	7	9	5	4

9. Find the means. Who produced more?

   (A) Ann, 6
   (B) Ann, 5.7
   (C) Bob, 5.7
   (D) Bob, 6

10. What is the mode for Ann's numbers?

    (A) 4
    (B) 5
    (C) 6
    (D) none

11. What is the mean absolute deviation of $\{2, 2, 3, 4, 14\}$?

    (A) 5
    (B) 3.6
    (C) 2.8
    (D) 3.0

12. Oxford shirt company wants to test how well their new button attachment machine has been working. They decide to test every tenth shirt that comes off the machine for an hour on Tuesday afternoon. Which of the following is a true statement regarding their sampling technique?

    (A) The Oxford shirt company describes a dependent event.
    (B) The Oxford shirt company describes an unbiased event.
    (C) The event is a simple random sample.
    (D) The sampling technique describes a representative sample.

13. A high school principal wants to survey school parents about requiring school uniforms next year. His school has 2700 families. He cuts up a calendar and puts all the dates from this year in a box. He then has his secretary mix up the papers and draw out 70 dates. He then sends out a survey to the parents of the students that were born on the dates that were selected. Which of the following is a true statement regarding his sampling technique?

    (A) The sampling technique describes a dependent sample.
    (B) The sampling technique describes a biased sample.
    (C) The event described is a simple random sample.
    (D) The sampling technique describes an unrepresentative sample.

Chapter 13
More Functions

This chapter covers the following Georgia Performance Standards:

| MM1A | Algebra | MM1A1h |
| | | MM1A1i |

13.1 Function Symmetry

In Mathematics, functions can be defined as symmetrical with respect to the y-axis or the origin. To test equations for symmetry, it is helpful to remember the following:

$f(-x) = f(x)$ means the function is symmetrical with respect to the y-axis.
Being symmetrical with the y-axis means the function is **even**.

$f(-x) = -f(x)$ means the function is symmetrical with respect to the origin.
Being symmetrical with the origin means the function is **odd**.

$f(-x) \neq -f(x)$ or $f(x)$ then the function is not symmetrical.

Example 1: Test the following function for symmetry. $f(x) = x^4 + x^2 + 3$

  Step 1: First, we need to substitute $-x$ in the function for x.
  $f(-x) = (-x)^4 + (-x)^2 + 3$

  Step 2: Carry out the operations.
  $f(-x) = x^4 + x^2 + 3$

  Step 3: Since $f(-x) = x^4 + x^2 + 3$ and $f(x) = x^4 + x^2 + 3$, then $f(-x) = f(x)$.
  This means $f(x) = x^4 + x^2 + 3$ is symmetrical with respect to the y-axis.
  The function $f(x) = x^4 + x^2 + 3$ is even.

Example 2:   Test the following function for symmetry. $f(x) = x^5 + x^3 + x$

Step 1:   First, we need to substitute $-x$ in the function for x.
$$f(-x) = (-x)^5 + (-x)^3 + (-x)$$

Step 2:   Carry out the operations.
$$f(-x) = -x^5 - x^3 - x$$

Step 3:   Factor out a negative. $f(-x) = -x^5 - x^3 - x = -(x^5 + x^3 + x)$. Since $f(-x) = -(x^5 + x^3 + x)$ and $-f(x) = -(x^5 + x^3 + x)$, then $f(-x) = -f(x)$. This means $f(x) = x^5 + x^3 + x$ is symmetrical with respect to the origin. The function $f(x) = x^5 + x^3 + x$ is odd.

Example 3:   Test the following function for symmetry. $f(x) = x^7 + x^4 - x^2$

Step 1:   First, we need to substitute $-x$ in the function for x.
$$f(-x) = (-x)^7 + (-x)^4 - (-x)^2$$

Step 2:   Carry out the operations.
$$f(-x) = -x^7 + x^4 - x^2$$

Step 3:   In this case, we see that $f(-x) \neq -f(x)$ or $f(x)$, which means this function is neither even nor odd (no symmetry).

Determine whether the function is even, odd, or neither.

1. $f(x) = x^3 + x^2 + x + 1$

2. $f(x) = 2x^3 - 4x$

3. $f(x) = 7x^2 - 11$

4. $f(x) = 8x^2 + x^4$

5. $f(x) = 4x^3 + x$

6. $f(x) = 6x^3 + 2x^2 - x$

7. $f(x) = x^5 + x^3 + 11$

8. $f(x) = x^7 - x^4 + x^2 - 11$

13.2   Symmetry of Graphs of Functions

Just like the symmetry of functions was determined algebraically, it can also be determined by looking at the graph. If you are unable to tell just by looking at the graph, you can check your answers using the following

y-axis (even)	If (a, b) is on the graph, so is $(-a, b)$.
origin (odd)	If (a, b) is on the graph, so is $(-a, -b)$.
x-axis	If (a, b) is on the graph, so is $(a, -b)$.

Example 4:   Determine if the graph of $y = x^2$ is symmetrical. If it is, is it odd, even, or neither?

Step 1:   Graph the function.

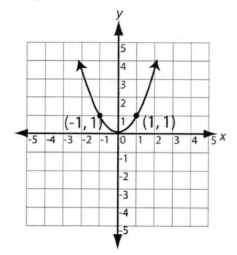

Looking at the graph we see that the graph is symmetrical about the y-axis. In other words, whatever the graph shows on one side of the y-axis, it is the same on the other side of the y-axis.

Step 2:   We check using the points on the graph. We will use the point $(1, 1)$ as it is a solution to the equation.
According to the table above, if $(1, 1)$ is a solution to a graph that is symmetrical about the y-axis, then $(-1, 1)$ must also be a solution.
Is it? $y = (-1)^2 = 1$. Yes, it is.
We also know since the graph is symmetrical about the y-axis that the graph is even.

Example 5:   Determine if the graph of $x = y^2$ is symmetrical? If it is, is it odd, even, or neither?

Step 1:   Graph the function.

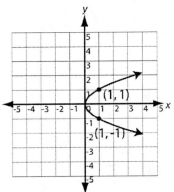

Looking at the graph we see that the graph is symmetrical about the x-axis. In other words, whatever the graph shows on one side of the x-axis, it is the same on the other side of the x-axis.

Step 2:   We check using the points on the graph. We will use the point $(1, 1)$ as it is a solution to the equation. If $(1, 1)$ is a solution to a graph that is symmetrical about the x-axis, then $(1, -1)$ must also be a solution. Is it? Yes.
We also know since the graph is symmetrical about the x-axis that the graph is neither even or odd.

Example 6:   Determine if the graph of $y = \dfrac{1}{x}$ is symmetrical? If it is, is it odd, even, or neither?

Step 1:   Graph the function.

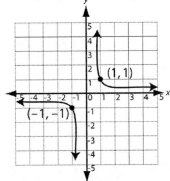

Looking at the graph we see that the graph is symmetrical about the origin. In other words, if you rotate one part of the graph $180°$ around the origin, it will be the exact same as the other half of the graph.

Step 2:   We check using the points on the graph. We will use the point $(1, 1)$ as it is a solution to the equation. If $(1, 1)$ is a solution to a graph that is symmetrical about the origin, then $(-1, -1)$ must also be a solution. Is it? Yes.
We also know since the graph is symmetrical about the origin that the graph is odd.

Tell whether the graph is symmetrical with respect to the x-axis, y-axis, the origin, or neither. Also, tell whether the graph is odd, even, or neither.

1.

2.

3.

4.

5.

6.
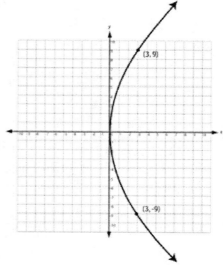

13.3   Finding Common Solutions for Intersecting Lines

When two lines intersect, they share exactly one point in common.

Example 7: $3x + 4y = 20$ and $2y - 4x = 12$

Put each equation in slope-intercept form.

$$
\begin{array}{ll}
3x + 4y = 20 & \qquad 2y - 4x = 12 \\
4y = -3x + 20 & \qquad 2y = 4x + 12 \\
y = -\frac{3}{4}x + 5 & \qquad y = 2x + 6
\end{array}
$$

slope-intercept form

Straight lines with different slopes are **intersecting lines**. Look at the graphs of the lines on the same Cartesian plane.

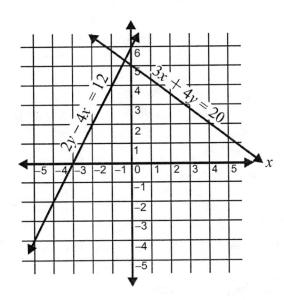

You can see from looking at the graph that the intersecting lines share one point in common.

13.4   Solutions of Equations

An intersection point is a point where two functions meet. To find the intersection point, you must set the equations equal to each other, $f(x) = g(x)$.

Example 8:   Find the intersection point(s) of $f(x) = x + 5$ and $g(x) = 2x + 6$.

   Step 1:   Find the value of x in the intersection point by setting the two equations equal to each other.

$$
\begin{aligned}
x + 5 &= 2x + 6 &&\text{Set the equations equal to each other.} \\
5 &= x + 6 &&\text{Subtract } x \text{ from both sides and simplify.} \\
-1 &= x &&\text{Subtract 6 from both sides and simplify.}
\end{aligned}
$$

   Step 2:   We use the value of x to find the intersection point by substituting the x-value back into one of the original equations.

$$
\begin{aligned}
f(x) &= x + 5 &&\text{Choose an equation.} \\
f(-1) &= -1 + 5 &&\text{Substitute the } x\text{-value into the equation.} \\
f(-1) &= 4 &&\text{Simplify.}
\end{aligned}
$$

   The intersection point is $(-1, 4)$.

Example 9:   Find the intersection point(s) of $f(x) = x^2$ and $g(x) = -3x + 4$.

   Step 1:   Find the value of x in the intersection point by setting the two equations equal to each other.

$$
\begin{aligned}
x^2 &= -3x + 4 &&\text{Set the equations equal to each other.} \\
x^2 + 3x - 4 &= 0 &&\text{Move all terms to one side the equation.} \\
(x + 4)(x - 1) &= 0 &&\text{Factor.} \\
x = -4 \text{ or } x &= 1 &&\text{Solve for } x.
\end{aligned}
$$

   Step 2:   We use the values of x to find the intersection points by substituting the x-values back into one of the original equations.

$$
\begin{aligned}
f(-4) &= (-4)^2 &&\text{Substitute one } x\text{-value into an original equation.} \\
f(-4) &= 16 &&\text{Solve.}
\end{aligned}
$$

$$
\begin{aligned}
f(1) &= 1^2 &&\text{Substitute the other } x\text{-value to find the 2nd intersection point.} \\
f(1) &= 1 &&\text{Solve.}
\end{aligned}
$$

   The intersection points are $(-4, 16)$ and $(1, 1)$.

Find the intersection points of the pairs of functions.

1. $f(x) = 2x + 3$ and $g(x) = -\frac{1}{2}x + 7$

2. $f(x) = 2x - 3$ and $g(x) = x + 4$

3. $f(x) = 5x - 1$ and $g(x) = 2x + 8$

4. $f(x) = x^2 + 6$ and $g(x) = -5x$

5. $f(x) = x^2 - x$ and $g(x) = 4x - 6$

6. $f(x) = x^2 - 3x$ and $g(x) = -4x + 6$

7. $f(x) = x^2 + 4x$ and $g(x) = 7x + 18$

8. $f(x) = 4x + 4$ and $g(x) = 16x + 4$

Chapter 13 Review

Solve the following problems about function symmetry.

1. True or False. If for a function $f(x)$, $f(-x) = -f(x)$ is true, then the function is odd.

2. True or False. If for a function $f(x)$, $f(-x) = f(x)$ is true, then the function is even.

3. Fill in the blank. When all exponents in a function are ____, the function will be even.

4. Is the function $f(x) = x^6 + x^4 + 7$ even, odd, or neither?

5. Is the function $f(x) = x^5 + x^3 + 2$ even, odd, or neither?

6. Fill in the blank. When all exponents in a function are ____, the function will be odd.

7. Is the function $f(x) = x^4 - x + 2$ even, odd, or neither?

State whether the graphs are symmetrical with respect to the x-axis, y-axis, origin, or neither.

8.

9.

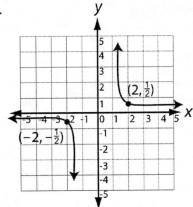

Find the intersection points of the pairs of functions.

10. $f(x) = x^2 - 7x - 8$ and $g(x) = 7x + 7$

11. $f(x) = 7x - 8$ and $g(x) = 5x + 2$

12. $f(x) = x^2 + 2x + 1$ and $g(x) = -2x^2 - 3x - 1$

Chapter 13 Test

1. What is the intersection point of the following functions?

   $f(x) = 4x - 8$ and $g(x) = 2x$

   (A) $(-4, -8)$
   (B) $(4, 8)$
   (C) $(-1, -2)$
   (D) $(1, 2)$

2. Is the following function even, odd, neither, or not enough information?
   $f(x) = x^{10} + x^6 + x^2 + 4$

   (A) even
   (B) odd
   (C) neither
   (D) not enough information

3. Is the following function even, odd, neither, or not enough information?
   $f(x) = x^5 + x^4 + x^2 + x + 4$

   (A) even
   (B) odd
   (C) neither
   (D) not enough information

4. Is the following function even, odd, neither, or not enough information?
   $f(x) = 2x^3 - 7x$

   (A) even
   (B) odd
   (C) neither
   (D) not enough information

5. Is the following function symmetrical?
   $f(x) = 6x^2 - 13$

   (A) yes, about the y-axis
   (B) yes, about the x-axis
   (C) yes, about the origin
   (D) no

6. Is the following function symmetrical?
   $f(x) = 7x^5 - 4x - 1$

   (A) yes, about the y-axis
   (B) yes, about the x-axis
   (C) yes, about the origin
   (D) no

7. Is the following graph symmetrical?

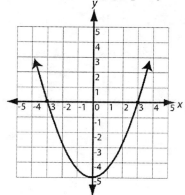

   (A) yes, about the y-axis
   (B) yes, about the x-axis
   (C) yes, about the origin
   (D) no

8. How many times do $f(x) = x + 2$ and $g(x) = 2x + 4$ intersect?

   (A) 0
   (B) 1
   (C) 2
   (D) 3

9. How many times do $f(x) = x^2$ and $g(x) = -x^2$ intersect?

   (A) 0
   (B) 1
   (C) 2
   (D) 3

10. What is the intersection point of $f(x) = -x$ and $g(x) = 2x + 1$?

(A) $\left(\frac{1}{3}, -\frac{1}{3}\right)$

(B) $\left(-\frac{1}{3}, 1\right)$

(C) $\left(-\frac{1}{3}, \frac{1}{3}\right)$

(D) $\left(-\frac{1}{3}, \frac{2}{3}\right)$

11. What is the intersection point of $f(x) = 2x + 1$ and $g(x) = x + 5$?

(A) $(2, 5)$

(B) $\left(\frac{3}{4}, \frac{5}{2}\right)$

(C) $(3, 8)$

(D) $(4, 9)$

12. Where do $f(x) = x^2 - 2x$ and $g(x) = 3x - 6$ intersect?

(A) $(2, 0)$ and $(3, 3)$

(B) $(2, 3)$ and $(3, 0)$

(C) $(2, 1)$ and $(3, 1)$

(D) $(2, 3)$ and $(3, 2)$

Chapter 14
Solving Equations

This chapter covers the following Georgia Performance Standards:

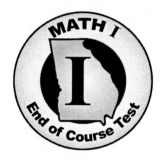

MM1A	Algebra	MM1A3b
		MM1A3c
		MM1A3d

14.1 Two-Step Algebra Problems

In the following two-step algebra problems, **addition** and **subtraction** are performed first and then **multiplication** and **division**.

Example 1: $-4x + 7 = 31$

    Step 1: Subtract 7 from both sides.

$$\begin{array}{rl} -4x + 7 & = 31 \\ -7 & -7 \\ \hline -4x & = 24 \end{array}$$

    Step 2: Divide both sides by -4.

$$\frac{\cancel{-4}x}{\cancel{-4}} = \frac{24}{-4} \qquad \text{so} \qquad x = \frac{24}{-4} = -6$$

Example 2: $-8 - y = 12$

    Step 1: Add 8 to both sides.

$$\begin{array}{rl} -8 - y & = 12 \\ +8 & +8 \\ \hline -y & = 20 \end{array}$$

    Step 2: To finish solving a problem with a negative sign in front of the variable, multiply both sides by -1. The variable cannot have a negative sign in front of it.

$$(-1)(-y) = (-1)(20) \qquad \text{so} \qquad y = -20$$

Solve the two-step algebra problems below.

1. $6x - 4 = -34$

2. $5y - 3 = 32$

3. $8 - t = 1$

4. $10p - 6 = -36$

5. $11 - 9m = -70$

6. $4x - 12 = 24$

7. $3x - 17 = -41$

8. $9d - 5 = 49$

9. $10h + 8 = 78$

10. $-6b - 8 = 10$

11. $-g - 24 = -17$

12. $-7k - 12 = 30$

13. $9 - 5r = 64$

14. $6y - 14 = 34$

15. $12f + 15 = 51$

16. $21t + 17 = 80$

17. $20y + 9 = 149$

18. $15p - 27 = 33$

19. $22h + 9 = 97$

20. $-5 + 36w = 175$

14.2   Two-Step Algebra Problems with Fractions

An algebra problem may contain a fraction. Study the following example to understand how to solve algebra problems that contain a fraction.

Example 3: $\dfrac{x}{2} + 4 = 3$

Step 1:
$$\dfrac{x}{2} + 4 = 3$$
$$\underline{-4 \qquad -4}$$
$$\dfrac{x}{2} \qquad = -1$$
Subtract 4 from both sides.

Step 2: $\dfrac{x}{2} = -1$ Multiply both sides by 2 to eliminate the fraction.

$$\dfrac{x}{\cancel{2}} \times \cancel{2} = -1 \times 2, \ x = -2$$

Simplify the following algebra problems.

1. $4 + \dfrac{y}{3} = 7$

2. $\dfrac{a}{2} + 5 = 12$

3. $\dfrac{w}{5} - 3 = 6$

4. $\dfrac{x}{9} - 9 = -5$

5. $\dfrac{b}{6} + 2 = -4$

6. $7 + \dfrac{z}{2} = -13$

7. $\dfrac{x}{2} - 7 = 3$

8. $\dfrac{c}{5} + 6 = -2$

9. $3 + \dfrac{x}{11} = 7$

10. $16 + \dfrac{m}{6} = 14$

11. $\dfrac{p}{3} + 5 = -2$

12. $\dfrac{t}{8} + 9 = 3$

13. $\dfrac{v}{7} - 8 = -1$

14. $5 + \dfrac{h}{10} = 8$

15. $\dfrac{k}{7} - 9 = 1$

16. $\dfrac{y}{4} + 13 = 8$

17. $15 + \dfrac{z}{14} = 13$

18. $\dfrac{b}{6} - 9 = -14$

19. $\dfrac{d}{3} + 7 = 12$

20. $10 + \dfrac{b}{6} = 4$

21. $2 + \dfrac{p}{4} = -6$

22. $\dfrac{t}{7} - 9 = -5$

23. $\dfrac{a}{10} - 1 = 3$

24. $\dfrac{a}{8} + 16 = 9$

14.3   Combining Like Terms

In an algebra problem, **terms** are separated by $+$ and $-$ signs. The expression $5x - 4 - 3x + 7$ has 4 terms: $5x$, 4, $3x$, and 7. Terms having the same variable can be combined (added or subtracted) to simplify the expression. $5x - 4 - 3x + 7$ simplifies to $2x + 3$.

$$5x - 3x \quad - 4 + 7 \; = 2x + 3$$

Simplify the following expressions.

1. $7x + 12x$

2. $8y - 5y + 8$

3. $4 - 2x + 9$

4. $11a - 16 - a$

5. $9w + 3w + 3$

6. $-5x + x + 2x$

7. $w - 15 + 9w$

8. $21 - 10t + 9 - 2t$

9. $-3 + x - 4x + 9$

10. $7b + 12 + 4b$

11. $4h - h + 2 - 5$

12. $-6k + 10 - 4k$

13. $2a + 12a - 5 + a$

14. $5 + 9c - 10$

15. $-d + 1 + 2d - 4$

16. $-8 + 4h + 1 - h$

17. $12x - 4x + 7$

18. $10 + 3z + z - 5$

19. $14 + 3y - y - 2$

20. $11p - 4 + p$

21. $11m + 2 - m + 1$

14.4   Solving Equations with Like Terms

When an equation has two or more like terms on the same side of the equation, combining like terms is the **first** step in solving the equation.

Example 4: $7x + 2x - 7 = 21 + 8$

Step 1:   Combine like terms on both sides of the equation.

Step 2:   Solve the two-step algebra problem as explained previously.

$$
\begin{aligned}
7x + 2x - 7 &= 21 + 8 \\
9x - 7 &= 29 \\
+7 \quad &\quad +7 \\
9x \div 9 &= 36 \div 9 \\
x &= 4
\end{aligned}
$$

Solve the equations below combining like terms first.

1. $3w - 2w + 4 = 6$

2. $7x + 3 + x = 16 + 3$

3. $5 - 6y + 9y = -15 + 5$

4. $-14 + 7a + 2a = -5$

5. $-2t + 4t - 7 = 9$

6. $9d + d - 3d = 14$

7. $-6c - 4 - 5c = 10 + 8$

8. $15m - 9 - 6m = 9$

9. $-4 - 3x - x = -16$

10. $9 - 12p + 5p = 14 + 2$

11. $10y + 4 - 7y = -17$

12. $-8a - 15 - 4a = 9$

If the equation has like terms on both sides of the equation, you must get all of the terms with a **variable** on one side of the equation and all of the **integers** on the other side of the equation.

Example 5: $3x + 2 = 6x - 1$

Step 1:	Subtract $6x$ from both sides to have all the **variables** on the left side.
Step 2:	Subtract 2 from both sides to have all the **integers** on the right side.
Step 3:	Divide by -3 to solve for x.

$$3x + 2 = 6x - 1$$
$$\underline{-6x \qquad\quad -6x}$$
$$-3x + 2 = -1$$
$$\underline{-2 \qquad -2}$$
$$\frac{-3x}{-3} = \frac{-3}{-3}$$
$$x = 1$$

Solve the following problems.

1. $3a + 1 = a + 9$

2. $2d - 12 = d + 3$

3. $5x + 6 = 14 - 3x$

4. $15 - 4y = 2y - 3$

5. $9w - 7 = 12w - 13$

6. $10b + 19 = 4b - 5$

7. $-7m + 9 = 29 - 2m$

8. $5x - 26 = 13x - 2$

9. $19 - p = 3p - 9$

10. $-7p - 14 = -2p + 11$

11. $16y + 12 = 9y + 33$

12. $13 - 11w = 3 - w$

13. $-17b + 23 = -4 - 8b$

14. $k + 5 = 20 - 2k$

15. $12 + m = 4m + 21$

16. $7p - 30 = p + 6$

17. $19 - 13z = 9 - 12z$

18. $8y - 2 = 4y + 22$

19. $5 + 16w = 6w - 45$

20. $-27 - 7x = 2x + 18$

21. $-12x + 14 = 8x - 46$

22. $27 - 11h = 5 - 9h$

23. $5t + 36 = -6 - 2t$

24. $17y + 42 = 10y + 7$

25. $22x - 24 = 14x - 8$

26. $p - 1 = 4p + 17$

27. $4d + 14 = 3d - 1$

28. $7w - 5 = 8w + 12$

29. $-3y - 2 = 9y + 22$

30. $17 - 9m = m - 23$

14.5   Multi-Step Algebra Problems

You can now use what you know about removing parentheses, combining like terms, and solving simple algebra problems to solve problems that involve three or more steps. Study the examples below to see how easy it is to solve multi-step problems.

Example 6: $3(x+6) = 5x - 2$

Step 1:	Use the distributive property to remove parentheses.	$3x + 18 = 5x - 2$
Step 2:	Subtract $5x$ from each side to have the terms with variables on the left side of the equation.	$\dfrac{-5x \qquad\qquad -5x}{-2x + 18 = -2}$
Step 3:	Subtract 18 from each side to have the integers on the right side of the equation.	$\dfrac{-18 \qquad -18}{\dfrac{-2x}{-2} = \dfrac{-20}{-2}}$
Step 4:	Divide both sides by -2 to solve for x.	$x = 10$

Example 7: $\dfrac{3(x-3)}{2} = 9$

Step 1:	Use the distributive property to remove parentheses.	$\dfrac{3x - 9}{2} = 9$
Step 2:	Multiply both sides by 2 to eliminate the fraction.	$\dfrac{2(3x-9)}{2} = 2(9)$
Step 3:	Add 9 to both sides, and combine like terms.	$3x - 9 = 18$
		$\dfrac{+9 \qquad +9}{\dfrac{3x}{3} = \dfrac{27}{3}}$
Step 4:	Divide both sides by 3 to solve for x.	$x = 9$

Solve the following multi-step algebra problems.

1. $2(y-3) = 4y + 6$

2. $\dfrac{2(a+4)}{2} = 12$

3. $\dfrac{10(x-2)}{5} = 14$

4. $\dfrac{12y - 18}{6} = 4y + 3$

5. $2x + 3x = 30 - x$

6. $\dfrac{2a+1}{3} = a + 5$

7. $5(b-4) = 10b + 5$

8. $-8(y+4) = 10y + 4$

9. $\dfrac{x+4}{-3} = 6 - x$

10. $\dfrac{4\,(n+3)}{5} = n - 3$

11. $3\,(2x - 5) = 8x - 9$

12. $7 - 10a = 9 - 9a$

13. $7 - 5x = 10 - (6x + 7)$

14. $4\,(x - 3) - x = x - 6$

15. $4a + 4 = 3a - 4$

16. $-3\,(x - 4) + 5 = -2x - 2$

17. $5b - 11 = 13 - b$

18. $\dfrac{-4x + 3}{2x} = \dfrac{7}{2x}$

19. $-(x + 1) = -2\,(5 - x)$

20. $4\,(2c + 3) - 7 = 13$

21. $6 - 3a = 9 - 2\,(2a + 5)$

22. $-5x + 9 = -3x + 11$

23. $3y + 2 - 2y - 5 = 4y + 3$

24. $3y - 10 = 4 - 4y$

25. $-(a + 3) = -2\,(2a + 1) - 7$

26. $5m - 2\,(m + 1) = m - 10$

27. $\dfrac{1}{2}\,(b - 2) = 5$

28. $-3\,(b - 4) = -2b$

29. $4x + 12 = -2\,(x + 3)$

30. $\dfrac{7x + 4}{3} = 2x - 1$

31. $9x - 5 = 8x - 7$

32. $7x - 5 = 4x + 10$

33. $\dfrac{4x + 8}{2} = 6$

34. $2\,(c + 4) + 8 = 10$

35. $y - (y + 3) = y + 6$

36. $4 + x - 2\,(x - 6) = 8$

14.6   Solving Equations Involving Radicals

Some multi-step equations contain radicals. An example of a radical is a square root, $\sqrt{}$. Work these type of equations out similarly to the section on equations.

Example 8:   Solve the following equation for x. $\sqrt{4x - 3} + 2 = 5$

Step 1:   The first step is to get the constants that are not under the radical on one side. Subtract 2 from both sides of the equation.
$$\sqrt{4x - 3} + 2 - 2 = 5 - 2$$
$$\sqrt{4x - 3} + 0 = 3$$
$$\sqrt{4x - 3} = 3$$

Step 2:   Next, you must eliminate the radical sign by squaring both sides of the equation.
$$\left(\sqrt{4x - 3}\right)^2 = (3)^2$$
$$4x - 3 = 9$$

Step 3:   Add 3 to both sides of the equation to get the constants on just one side of the equation.
$$4x - 3 + 3 = 9 + 3$$
$$4x + 0 = 12$$
$$4x = 12$$

Step 4:   Last, get x on one side of the equation by itself by dividing both sides by 4.
$$\frac{4x}{4} = \frac{12}{4}$$
$$x = 3$$

Solve the following equations.

1. $\sqrt{x} - 13 = -8$

2. $3 + \sqrt{7t - 3} = 5$

3. $\sqrt{3q + 12} - 4 = 5$

4. $\sqrt{f} + 2 = 8$

5. $5 = \sqrt{6g - 5} + (-2)$

6. $2 = \sqrt{x - 3}$

7. $\sqrt{t - 3} = 1$

8. $\sqrt{-d + 1} - 9 = -6$

9. $10 - \sqrt{x} = 9$

10. $\sqrt{15y + 4} + 4 = 12$

11. $\sqrt{r + 14} - 1 = 9$

12. $3 - \sqrt{2q - 1} = 0$

13. $\sqrt{t} + 4 = 13$

14. $17 = \sqrt{23 - f} + 15$

15. $19 - \sqrt{7x - 5} = 16$

14.7   Finding the Intercepts of a Quadratic Equation

A **quadratic equation** is an equation where either the y or x variable is squared. Finding the intercepts of a quadratic equation is similar to finding the intercepts of a line. In most cases, the variable x is squared, which means there could be two x-intercepts. There could be one, two, or zero x-intercepts. The x-intercepts of a quadratic equation are called the **zeros** of a quadratic equation.

Example 9:   Find the x-intercept(s) of the quadratic equation, $y = x^2 - 4$.

Step 1:   Find the x-intercept(s). The x-intercept is the point, or points in this case, where the graph crosses the x-axis. In this case, we plug 0 in for y because y is always zero along the x-axis. Solve for x.

$$y = x^2 - 4$$
$$0 = x^2 - 4$$
$$0 + 4 = x^2 - 4 + 4$$
$$4 = x^2$$
$$\sqrt{4} = \sqrt{x^2}$$
$$\sqrt{4} = x$$

Thus, $x = \sqrt{4}$ and $\sqrt{4} = -2$ or 2, so $x = -2$ or $x = 2$.
The x-intercepts are $(-2, 0)$ and $(2, 0)$.

Step 2:   To verify that the intercepts are correct, graph the equation on the coordinate plane.

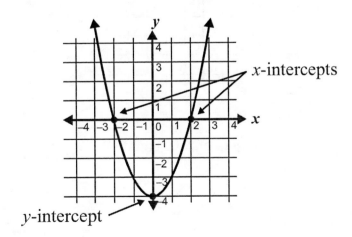

x-intercepts

y-intercept

Example 10:   Find the x-intercept(s) of the quadratic equation, $y = x^2 + 2$.

Step 1:   Find the x-intercept. Plug 0 in for y and solve for x.

$y = x^2 + 2$
$0 = x^2 + 2$
$0 - 2 = x^2 + 2 - 2$
$-2 = x^2$
$\sqrt{-2} = \sqrt{x^2}$
$\sqrt{-2} = x$

You cannot take the square root of a negative number and get a real number as the answer, so there is no x-intercept for this quadratic equation.

Step 3:   To verify that the intercepts are correct, graph the equation on the coordinate plane. As you can see from the graph, the tails of the parabola are increasing in the positive y direction, so they will never come back down, which means they will never cross the x-axis.

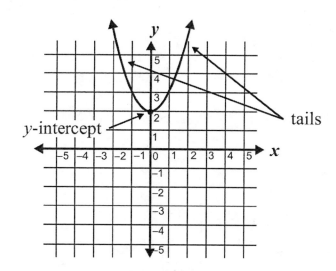

Find the x-intercept(s) of the following quadratic equations.

1. $y = x^2$

2. $y = 2x^2 - 4$

3. $y = -x^2 + 2$

4. $y = x^2 + 6$

5. $y = -x^2 - 1$

6. $y = x^2 - 5$

7. $y = 4x^2 + 8$

8. $y = x^2 + 2x + 1$

9. $y = x^2 - 7x + 12$

14.8   Solving Quadratic Equations Using Tables

A quadratic equation is an equation in the form of $y = ax^2 + bx + c$, where a, b, and c are real numbers and $a \neq 0$. The graph of a quadratic equation is a parabola.

Example 11:   Find the solution(s) to the function $f(x) = x^2 + 2x + 1$.

Step 1:   Make a table of ordered pairs that satisfy the equation. Then graph the ordered pairs on a coordinate plane.

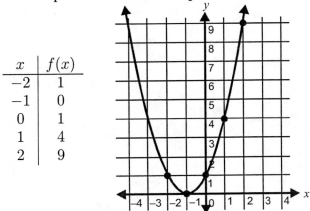

x	$f(x)$
-2	1
-1	0
0	1
1	4
2	9

Step 2:   We can see from step 1 that there are an infinite number of ordered pairs that lie on the parabola. What we want to find is where the parabola crosses the x-axis. This is known as finding the zeros of the equation (or the solutions to the equation). From the table we can see that when $f(x)$ is 0, x is -1. The x-intercept is -1. There is only one solution, $x = -1$.

Some quadratic equations cross the x-axis in two places, instead of one, like example 11. See example 12.

Example 12:   Find the solution(s) to the function $f(x) = x^2 + 5x + 6$.

Step 1:   Make a table of ordered pairs that satisfy the equation. Then graph the ordered pairs on a coordinate plane.

x	$f(x)$
-4	2
-3	0
-2	0
-1	2
0	6

Step 2:   From the table we can see that the parabola crosses the x-axis in two places, when x is -3 and -2. There are two solutions, $x = -3, -2$.

Look at the tables to find the solution(s). (Hint: The solutions are the x-intercepts.)

1. $f(x) = x^2 - 5x + 6$

x	$f(x)$
-1	12
0	6
1	2
2	0
3	0
4	2

4. $f(x) = x^2 + 4x + 3$

x	$f(x)$
0	3
-1	0
-2	-1
-3	0
-4	3
-5	8

2. $f(x) = x^2 - x - 20$

x	$f(x)$
-5	10
-4	0
0	-20
4	-8
5	0

5. $f(x) = -x^2$

x	$f(x)$
-2	-4
-1	-1
0	0
1	-1
2	-4

3. $f(x) = -x^2 + 4$

x	$f(x)$
-4	-12
-3	-5
-2	0
-1	3
0	4
1	3
2	0
3	-5

6. $f(x) = x^2 - x - 12$

x	$f(x)$
-3	0
-2	-6
-1	-10
0	-12
1	-12
2	-10
3	-6
4	0

14.9 Using the Quadratic Formula

You may be asked to use the quadratic formula to solve an algebra problem known as a **quadratic equation**. The equation should be in the form $ax^2 + bx + c = 0$.

Example 13: Using the quadratic formula, find x in the following equation: $x^2 - 8x = -7$.

Step 1: Make sure the equation is set equal to 0.

$$x^2 - 8x + 7 = -7 + 7$$
$$x^2 - 8x + 7 = 0$$

The quadratic formula, $\dfrac{-b \pm \sqrt{b^2 - 4ac}}{2a}$, will be given to you on your formula sheet with your test.

Step 2: In the formula, a is the number x^2 is multiplied by, b is the number x is multiplied by and c is the last term of the equation. For the equation in the example, $x^2 - 8x + 7$, $a = 1$, $b = -8$, and $c = 7$. When we look at the formula we notice a $\pm$ sign. This means that there will be two solutions to the equation, one when we use the plus sign and one when we use the minus sign. Substituting the numbers from the problem into the formula, we have:

$$\frac{8 + \sqrt{8^2 - (4)(1)(7)}}{2(1)} = 7 \qquad \text{or} \qquad \frac{8 - \sqrt{8^2 - (4)(1)(7)}}{2(1)} = 1$$

The solutions are $\{7, 1\}$

For each of the following equations, use the quadratic formula to find two solutions.

1. $x^2 + x - 6 = 0$

2. $y^2 - 2y - 8 = 0$

3. $a^2 + 2a - 15 = 0$

4. $y^2 - 5y + 4 = 0$

5. $b^2 - 9b + 14 = 0$

6. $x^2 - 3x - 4 = 0$

7. $y^2 + y - 20 = 0$

8. $d^2 + 6d + 8 = 0$

9. $y^2 - 7y + 12 = 0$

10. $x^2 - 3x - 28 = 0$

11. $a^2 - 5a + 6 = 0$

12. $b^2 + 3b - 10 = 0$

13. $a^2 + 7a - 8 = 0$

14. $c^2 + 3c + 2 = 0$

15. $x^2 - x - 42 = 0$

16. $a^2 + 5a - 6 = 0$

17. $b^2 + 7b + 12 = 0$

18. $y^2 + y - 12 = 0$

19. $a^2 - 3a - 10 = 0$

20. $d^2 + 10d + 16 = 0$

21. $x^2 - 4x - 12 = 0$

14.10 Real-World Quadratic Equations

The most common real life situation that would use a quadratic equation is the motion of an object under the force of gravity. Two examples are a ball being kicked into the air or a rocket being shot into the air.

Example 14: A high school football player is practicing his field goal kicks. The equation below represents the height of the ball at a specific time.

$$s = -9t^2 + 45t$$

t = amount of time in seconds

s = height in feet

Question 1: Where will the ball be at 4 seconds?

Solution 1: Since there are only two variables, you will only need the value of one variable to solve the problem. Simply plug in the number 4 in place of the variable t and solve the equation as shown below.

$$s = -9\,(4)^2 + 45\,(4)$$

$$s = -9\,(16) + 180$$

$$s = -144 + 180$$

$$s = 36$$

At 4 seconds the ball will be 36 ft in the air.

Question 2: If the ball is 54 ft in the air, how much time has gone by?

Solution 2: This question is similar to the previous one, except that the given variable is different. This time you would be replacing s with 54 and then solve the equation.

$54 = -9t^2 + 45t$	Subtract 54 on both sides.
$0 = -9t^2 + 45t - 54$	Divide the entire equation by -9.
$0 = t^2 - 5t + 6$	Factor the equation.
$0 = (t - 3)(t - 2)$	Solve for t.
$t = 3 \qquad t = 2$	

For this question we got 2 answers. The ball is 54 ft in the air when 2 and 3 seconds have gone by.

Example 15:   John and Alex are kicking a soccer ball back and forth to each other. The equation below represents the height of the ball at a specific point in time.
$s = -4t^2 + 24t$, where $t =$ amount of time in seconds and $s =$ height in feet

Question 1:   How long does it take for the soccer ball to come back down to the ground?

Solution 1:   Looking at this problem you can see that no value was given, but one was indirectly given. The question asks when will the ball come back down. This is just another way of asking, "When will the height of the ball be zero?" The value 0 will be used for s. Substitute 0 back in for s and solve.

$0 = -4t^2 + 24t$ Factor out the greatest common factor.
$0 = -4t\,(t - 6)$ Set each factor to 0 and solve.
$t = 0$ $t = 6$

It is clear that the ball is on the ground at 0 seconds, so the value of $t = 0$ is not the answer and the second value of t is used instead. It takes the ball 6 seconds to go up into the air and then come back down to the ground.

Question 2:   What is the highest point the ball will go?

Solution 2:   This is asking what is the vertex of the equation. You will need to use the vertex formula. As a reminder, the quadratic equation is defined as $y = ax^2 + bx + c$, where $a \neq 0$. The quadratic equation can also be written as a function of x by substituting $f\,(x)$ for y, such as $f\,(x) = ax^2 + bx + c$. To find the point of the vertex of the graph, you must use the formula below.

$$\text{vertex} = \left(-\frac{b}{2a}, f\left(-\frac{b}{2a} \right) \right)$$

where $f\left(-\frac{b}{2a} \right)$ is the quadratic equation evaluated at the value $-\frac{b}{2a}$. To do this, plug $-\frac{b}{2a}$ in for x.

To use the vertex equation, put the original equation in quadratic form and find a and b.
$s = -4t^2 + 24t \Rightarrow -4t^2 + 24t + 0$.
Since a is the coefficient of t^2, $a = -4$. b is the coefficient of t, so $b = 24$.

Find the solution to $-\frac{b}{2a}$ by substituting the values of a and b from the equation into the expression.
$$-\frac{b}{2a} = -\left(\frac{24}{2 \times -4} \right) = -\left(\frac{24}{-8} \right) = -(-3) = 3$$

Find the solution to $f\left(-\frac{b}{2a} \right)$. We know that $-\frac{b}{2a} = 3$, so we need to find $f\,(3)$. To do this, we must substitute 3 into the quadratic equation for x.
$f\,(t) = -4t^2 + 24t$
$f\,(3) = -4\,(3)^2 + 24\,(3) = -4\,(9) + 72 = -36 + 72 = 36$

The vertex equals $(3, 36)$. This means at 3 seconds, the ball is 36 feet in the air. Therefore the highest the soccer ball will go is 36 ft.

213

Solve the following quadratic problems.

1. Eric is at the top of a cliff that is 500 ft from the ocean's surface. He is waiting for his friend to climb up and meet him. As he is waiting he decides to start casually tossing pebbles off the side of the cliff. The equation that represents the height of his pebbles tosses is $s = -t^2 + 5t + 500$, where s = distance in feet and t = time in seconds.

   (A) How long does it take the pebble to hit the water?

   (B) If fifteen seconds have gone by, what is the height of the pebble from the ocean?

   (C) What is the highest point the pebble will go?

2. Devin is practicing golf at the driving range. The equation that represents the height of his ball is $s = -0.5t^2 + 12t$, where s = distance in feet and t = time in seconds.

   (A) What is the highest her ball will ever go?

   (B) If the ball is at 31.5 ft in the air, how many seconds has gone by?

   (C) How long will it take for the ball to hit the ground?

3. Jack throws a ball up in the air to see how high he can get it to go. The equation that represents the height of the ball is $s = -2t^2 + 20t$, where s = distance in feet and t = time in seconds.

   (A) How high will the ball be at 7 seconds?

   (B) If the ball is 48 feet in the air, how many seconds have gone by?

   (C) How long does it take for the ball to go up and come back down to the ground?

4. Kali is jumping on her super trampoline, getting as high as she possibly can. The equation to represent her height is $s = -5t^2 + 20t$, where s = distance in feet and t = time in seconds.

   (A) What is the highest Kali can jump on her trampoline?

   (B) How high will Kali be at 4 seconds?

   (C) If Kali is 18.75 ft in the air, then how many seconds have gone by?

Chapter 14 Review

Solve each of the following equations.

1. $4a - 8 = 28$

2. $5 + \dfrac{x}{8} = -4$

3. $-7 + 23w = 108$

4. $\dfrac{y - 8}{6} = 7$

5. $c - 13 = 5$

6. $\dfrac{b + 9}{12} = -3$

7. $19 - 8d = d - 17$

8. $\dfrac{-3}{x} + 11 = -1$

9. $7w - 8w = -4w - 30$

10. $\dfrac{12}{f} - 7 = -5$

11. $6 + 16x = -2x - 12$

12. $6 - \dfrac{1}{q} = 4$

13. $3 - \sqrt{6 - 2x} = 4$

14. $\sqrt{w + 11} + 14 = 15$

15. $\sqrt{5k + 1} - 3 = 11$

16. $\dfrac{-11c - 35}{4} = 4c - 2$

17. $5 + x - 3(x + 4) = -17$

18. $\dfrac{5(n + 4)}{3} = n - 8$

Find the x-intercept(s) (solutions) of the following quadratic equations.

19. $y = 2x^2 - 8$

20. $y = 5x^2 + 1$

21. $f(x) = 2x^2 - 4x - 6$

x	$f(x)$
-2	10
-1	0
0	-6
1	-8
2	-6
3	0

22. $y = x^2 - 6$

23. $y = x^2 - x - 20$

24. $f(x) = x^2 + 4x + 3$

x	$f(x)$
-5	8
-4	3
-3	0
-2	-1
-1	0
0	3

25. $y = -x^2 - 9$

26. $f(x) = x^2 - 2x - 3$

27. $f(x) = x^2 + 5x + 6$

x	$f(x)$
-5	6
-4	2
-3	0
-2	0
-1	2
0	6

Chapter 14 Test

1. Find the value of n. $19n - 57 = 76$

    (A) 1
    (B) 3
    (C) 5
    (D) 7

2. $(-2, 1)$ is a solution for which of the following equations?

    (A) $y + 2x = 4$
    (B) $-2x - y = 5$
    (C) $x + 2y = -4$
    (D) $2x - y = -5$

3. What are the x-intercept(s) of the equation $y = -2x^2 + 8$?

    (A) $(-2, 0)$ and $(2, 0)$
    (B) $(-4, 0)$ and $(4, 0)$
    (C) $(8, 0)$
    (D) There are no x-intercepts.

4. Solve for x. $14x + 84 = 154$

    (A) 4
    (B) 5
    (C) 11
    (D) 17

5. Solve for x. $3(x - 2) - 1 = 6(x + 5)$

    (A) -4
    (B) $-\dfrac{37}{3}$
    (C) 4
    (D) $\dfrac{23}{3}$

6. What are the x-intercept(s) of the equation $y = x^2 + 9$

    (A) $(9, 0)$
    (B) $(-3, 0)$ and $(3, 0)$
    (C) $(-9, 0)$ and $(9, 0)$
    (D) There are no x-intercepts.

7. What are the solutions to $f(x) = x^2 + 4x - 5$?

    (A) $x = 1, -5$
    (B) $x = -1, 5$
    (C) $x = 1$
    (D) $x = 5$

8. What are the solutions to $f(x) = x^2 + 5x - 14$?

    (A) $x = -2$
    (B) $x = 2$
    (C) $x = -2, 7$
    (D) $x = 2, -7$

9. How many solutions would the equation represented by this chart have?

x	$f(x)$
-2	0
-1	1
0	3
1	1
2	0

    (A) 1
    (B) 2
    (C) 3
    (D) 0

Chapter 15
Coordinate Geometry

This chapter covers the following Georgia Performance Standards:

MM1G	Geometry	MM1G1a
		MM1G1b
		MM1G1c
		MM1G1d
		MM1G1e

15.1 Finding the Distance Between Two Points

Notice that a subscript added to the x and y identifies each ordered pair uniquely in the plane. For example, point 1 is identified as (x_1, y_1), point 2 as (x_2, y_2), and so on. This unique subscript identification allows us to calculate slope, distance, and midpoints of line segments in the plane using standard formulas like the distance formula. To find the distance between two points on a Cartesian plane, use the following formula:

$$d = \sqrt{(y_2 - y_1)^2 + (x_2 - x_1)^2}$$

Example 1:   Find the distance between $(-2, 1)$ and $(3, -4)$.

Plugging the values from the ordered pairs into the formula, we find:

$$d = \sqrt{(-4 - 1)^2 + [3 - (-2)]^2}$$
$$d = \sqrt{(-5)^2 + (5)^2}$$
$$d = \sqrt{25 + 25} = \sqrt{50}$$

To simplify, we look for perfect squares that are a factor of 50. $50 = 25 \times 2$. Therefore,
$$d = \sqrt{25} \times \sqrt{2} = 5\sqrt{2}$$

Find the distance between the following pairs of points using the distance formula above.

1. $(6, -1)\,(5, 2)$
2. $(-4, 3)\,(2, -1)$
3. $(10, 2)\,(6, -1)$
4. $(-2, 5)\,(-4, 3)$
5. $(8, -2)\,(3, -9)$

6. $(2, -2)\,(8, 1)$
7. $(3, 1)\,(5, 5)$
8. $(-2, -1)\,(3, 4)$
9. $(5, -3)\,(-1, -5)$
10. $(6, 5)\,(3, -4)$

11. $(-1, 0)\,(-9, -8)$
12. $(-2, 0)\,(-6, 6)$
13. $(2, 4)\,(8, 10)$
14. $(-10, -5)\,(2, -7)$
15. $(-3, 6)\,(1, -1)$

15.2   Finding the Distance Between a Point and a Line

The shortest distance between a point and a line is a perpendicular line segment.

Example 2:   Find the shortest distance between line l and point C.

Each unit on the grid equals 5 centimeters.

Step 1:   We know that since $\overline{CA}$ is perpendicular to line l, that the length of $\overline{CA}$ is the shortest distance between line l and point C.

Step 2:   To find the length of $\overline{CA}$, use the distance formula.
Find the coordinates of C and A. $C = (2,2)$ and $A = (-2,2)$
Plug the coordinates into the distance formula and solve.
$$d = \sqrt{(2-2)^2 + [(-2)-2]^2} = \sqrt{0 + (-4)^2} = \sqrt{16} = 4$$

Step 3:   The length of $\overline{CA}$ is 4 units and each unit represents 5 centimeters.
4 units × 5 centimeters = 20 centimeters
The shortest distance between line l and point C is 20 centimeters.

Find the shortest distance between the points and lines listed below. Use the graph to the right. Each unit equals 6 inches.

1.   line l and point G

2. $\overline{AB}$ and point H

3.   line l and point I

4. $\overline{ED}$ and point F

5. $\overline{AB}$ and point C

6.   line l and point E

7.   line l and point A

8. $\overline{AB}$ and point G

9.   line l and point H

10.   line l and point F

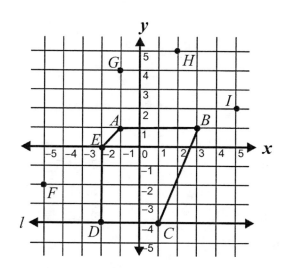

15.3   Finding the Midpoint of a Line Segment

You can use the coordinates of the endpoints of a line segment to find the coordinates of the midpoint of the line segment. The formula to find the midpoint between two coordinates is:

$$\text{midpoint, } M = \left(\frac{x_1 + x_2}{2}, \frac{y_1 + y_2}{2} \right)$$

Example 3:   Find the midpoint of the line segment having endpoints at $(-3, -1)$ and $(4, 3)$.
Use the formula for the midpoint. $M = \left(\dfrac{4 + (-3)}{2}, \dfrac{3 + (-1)}{2} \right)$
When we simplify each coordinate, we find the midpoint, M, is $\left(\frac{1}{2}, 1 \right)$.

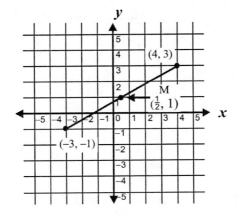

For each of the following pairs of points, find the coordinate of the midpoint, M, using the formula given above.

1. $(4, 5) \, (-6, 9)$

2. $(-3, 2) \, (-1, -2)$

3. $(3, 6) \, (9, 12)$

4. $(2, 5) \, (6, 9)$

5. $(8, 9) \, (6, 11)$

6. $(-4, 3) \, (8, 7)$

7. $(-1, -5) \, (-3, -11)$

8. $(4, 2) \, (-2, 8)$

9. $(4, 3) \, (-1, -5)$

10. $(-6, 2) \, (8, -8)$

11. $(-3, 9) \, (-9, 3)$

12. $(7, 8) \, (11, 6)$

13. $(12, 19) \, (2, 3)$

14. $(5, 4) \, (9, -2)$

15. $(-4, 6) \, (10, -2)$

15.4   Pythagorean Theorem

Pythagoras was a Greek mathematician and philosopher who lived around 600 B.C. He started a math club among Greek aristocrats called the Pythagoreans. Pythagoras formulated the **Pythagorean Theorem** which states that in a **right triangle**, the sum of the squares of the legs of the triangle are equal to the square of the hypotenuse. Most often you will see this formula written as $a^2 + b^2 = c^2$. **This relationship is only true for right triangles.**

Example 4:   Find the length of side c.

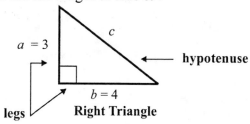

Formula:
$$a^2 + b^2 = c^2$$
$$3^2 + 4^2 = c^2$$
$$9 + 16 = c^2$$
$$25 = c^2$$
$$\sqrt{25} = \sqrt{c^2}$$
$$5 = c$$

Find the hypotenuse of the following triangles. Round the answers to two decimal places.

1.

$c =$ _____

4.

$c =$ _____

7.

$c =$ _____

2.

$c =$ _____

5.

$c =$ _____

8.

$c =$ _____

3.

$c =$ _____

6.

$c =$ _____

9.

$c =$ _____

15.5 Finding the Missing Leg of a Right Triangle

In some triangles, we know the measurement of the hypotenuse as well as one of the legs. To find the measurement of the other leg, use the Pythagorean theorem by filling in the known measurements, and then solve for the unknown side.

Example 5: Find the measure of b.

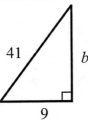

In the formula, $a^2 + b^2 = c^2$, a and b are the legs and c is always the hypotenuse.

$9^2 + b^2 = 41^2$

$81 + b^2 = 1681$

$b^2 = 1681 - 81$

$b^2 = 1600$

$\sqrt{b^2} = \sqrt{1600}$

$b = 40$

Practice finding the measure of the missing leg in each right triangle below. Simplify square roots.

1.

_____

4.

_____

7.

_____

2.

_____

5.

_____

8.

_____

3.

_____

6.

_____

9.

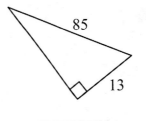

_____

15.6   Applications of the Pythagorean Theorem

The Pythagorean Theorem can be used to determine the distance between two points in some situations. Recall that the formula is written $a^2 + b^2 = c^2$.

Example 6:   Find the distance between point B and point A given that the length of each square is 1 inch long and 1 inch wide.

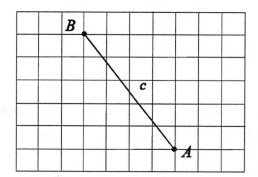

Step 1:   Draw a straight line between the two points. We will call this side c.

Step 2:   Draw two more lines, one from point B and one from point A. These lines should make a 90° angle. The two new lines will be labeled a and b. Now we can use the Pythagorean Theorem to find the distance from Point B to Point A.

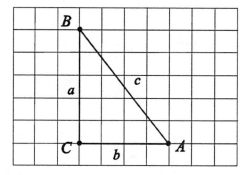

Step 3:   Find the length of a and c by counting the number of squares each line has. We find that $a = 5$ inches and $b = 4$ inches. Now, substitute the values found into the Pythagorean Theorem.

$$
\begin{aligned}
a^2 + b^2 &= c^2 \\
5^2 + 4^2 &= c^2 \\
25 + 16 &= c^2 \\
41 &= c^2 \\
\sqrt{41} &= \sqrt{c^2} \\
\sqrt{41} &= c
\end{aligned}
$$

Use the Pythagorean Theorem to find the distances asked. Round your answers to two decimal points.

Below is a diagram of the mall. Use the grid to help answer questions 1 and 2. Each square is 25 feet × 25 feet.

1. Marty walks from Pinky's Pet Store to the restroom to wash his hands. How far did he walk?

2. Betty needs to meet her friend at Silly Shoes, but she wants to get a hot dog first. If Betty is at Thrifty's, how far will she walk to meet her friend?

Below is a diagram of a football field. Use the grid on the football field to help find the answers to questions 3 and 4. Each square is 10 yards × 10 yards.

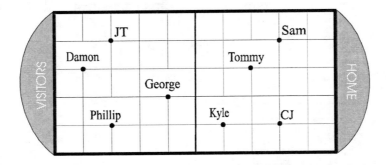

3. George must throw the football to a teammate before he is tackled. If CJ is the only person open, how far must George be able to throw the ball?

4. Damon has the football and is about to make a touchdown. If Phillip tries to stop him, how far must he run to reach Damon?

15.7   Drawing Geometric Figures on a Cartesian Coordinate Plane

You can use a Cartesian coordinate plane to draw geometric figures by plotting **vertices** and connecting them with line segments.

Example 7:   What are the coordinates of each vertex of quadrilateral $ABCD$ below?

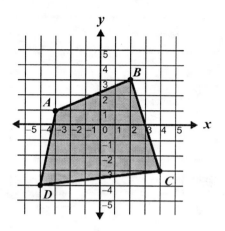

Step 1:   To find the coordinates of point A, count over -3 on the x-axis and up 1 on the y-axis. point $A = (-3, 1)$.

Step 2:   The coordinates of point B are located to the right two units on the x-axis and up 3 units on the y-axis. point $B = (2, 3)$.

Step 3:   Point C is located 4 units to the right on the x-axis and down -3 on the y-axis. point $C = (4, -3)$.

Step 4:   Point D is -4 units left on the x-axis and down -4 units on the y-axis. point $D = (-4, -4)$.

Example 8:   Plot the following points. Then construct and identify the geometric figure that you plotted.

$$A = (-2, -5) \qquad B = (-2, 1) \qquad C = (3, 1) \qquad D = (3, -5)$$

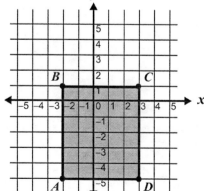

Figure $ABCD$ is a rectangle.

Find the coordinates of the geometric figures graphed below.

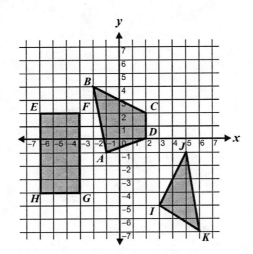

1. Quadrilateral $ABCD$

   $A =$ _____

   $B =$ _____

   $C =$ _____

   $D =$ _____

2. Rectangle $EFGH$

   $E =$ _____

   $F =$ _____

   $G =$ _____

   $H =$ _____

3. Triangle IJK

   $I =$ _____

   $J =$ _____

   $K =$ _____

4. Parallelogram $LMNO$

   $L =$ _____

   $M =$ _____

   $N =$ _____

   $O =$ _____

5. Right Triangle PQR

   $P =$ _____

   $Q =$ _____

   $R =$ _____

6. Pentagon $STVXY$

   $S =$ _____

   $T =$ _____

   $V =$ _____

   $X =$ _____

   $Y =$ _____

Plot and label the following points. Then construct and identify the geometric figure you plotted. Question 1 is done for you.

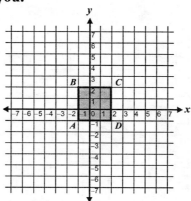

Figure   **Figure**

1. Point $A = (-1, -1)$
    Point $B = (-1, 2)$
    Point $C = (2, 2)$
    Point $D = (2, -1)$ **square**

3. Point $H = (-4, 0)$
    Point $I = (-6, 0)$
    Point $J = (-4, 4)$
    Point $K = (-2, 4)$ _____

2. Point $E = (3, -2)$
    Point $F = (5, 1)$
    Point $G = (7, -2)$ _____

4. Point $L = (-1, -3)$
    Point $M = (4, -6)$
    Point $N = (-1, -6)$ _____

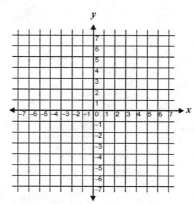

Figure   **Figure**

5. Point $A = (-2, -3)$
    Point $B = (-3, 5)$
    Point $C = (-1, 6)$
    Point $D = (1, 5)$
    Point $E = (0, 3)$ _____

7. Point $J = (-1, 2)$
    Point $K = (-1, -1)$
    Point $L = (3, -2)$ _____

6. Point $F = (-1, -3)$
    Point $G = (-3, -5)$
    Point $H = (-1, -7)$
    Point $I = (1, -5)$ _____

8. Point $M = (6, 2)$
    Point $N = (6, -4)$
    Point $O = (4, -4)$
    Point $P = (4, 2)$ _____

Chapter 15 Review

1. What is the distance between the points $(3, 3)$ and $(6, -1)$?

2. What is the distance between the two points $(-3, 0)$ and $(2, 5)$?

For questions 3–6, use the following graph.

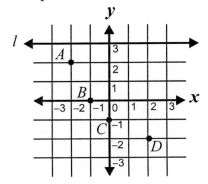

3. What is the shortest distance between line l and point A?

4. What is the shortest distance between line l and point B?

5. What is the shortest distance between line l and point C?

6. What is the shortest distance between line l and point D?

For questions 7 and 8, find the coordinates of the midpoint of the line segments with the given endpoints.

7. $(6, 10)$ $(-4, 4)$

8. $(-1, -7)$ $(5, 3)$

9. Find the missing side.

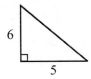

10. Find the measure of the missing leg of the right triangle below.

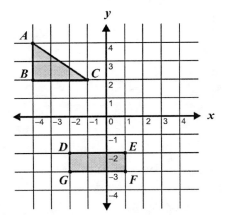

Find the coordinates of the geometric figures graphed above.

11. point A

12. point B

13. point C

14. point D

15. point E

16. point F

17. point G

Plot and label the following points on the same graph.

18. point $H = (1, 1)$

19. point $I = (3, 1)$

20. point $J = (4, -2)$

21. point $K = (2, -2)$

22. What type of figure did you plot?

Chapter 15 Test

1.

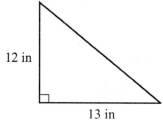

Logan enjoys taking his dog to the park. He walks down Hickory St., turns onto Maple Ave. to meet his friend, Brett, and then continues on Maple Ave. to the park. How far will Brett walk from his house to the park?

(A) 350 yards
(B) 437 yards
(C) 687 yards
(D) 532 yards

2. Approximately what is the measure of the hypotenuse of the triangle?

12 in

13 in

(A) 14 in
(B) 157 in
(C) 18 in
(D) 313 in

3. What is the measure of the missing side in the triangle?

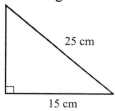

25 cm

15 cm

(A) 5 cm
(B) 29 cm
(C) 10 cm
(D) 20 cm

4. Find the midpoint of a line segment that has the endpoints $(-6, 2)$ and $(0, -2)$.

(A) $(0, -1)$
(B) $(-3, 4)$
(C) $(0, -4)$
(D) $(-3, 0)$

5. Randy plotted points G and H on the Cartesian coordinate graph below. Where could he plot points J and K if he wants to form a square $GHJK$?

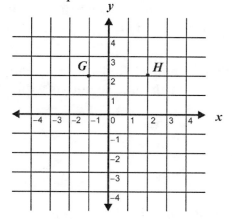

(A) J at $(2, 4)$ and K at $(-1, 4)$
(B) J at $(3, 0)$ and K at $(0, 0)$
(C) J at $(2, 0)$ and K at $(-1, 0)$
(D) J at $(2, -1)$ and K at $(-1, -1)$

Use the following coordinate grid to answer questions 6–8.

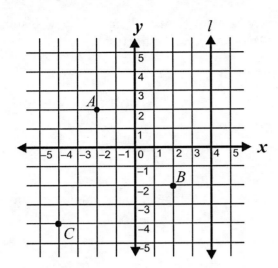

Each unit equals 9 millimeters.

6. What is the shortest distance between line l and point A?

(A) 6 millimeters
(B) 54 millimeters
(C) 60 millimeters
(D) 9 millimeters

7. What is the shortest distance between line l and point B?

(A) 18 millimeters
(B) 9 millimeters
(C) 2 millimeters
(D) 1 millimeter

8. What is the shortest distance between line l and point C?

(A) 108 millimeters
(B) 72 millimeters
(C) 54 millimeters
(D) 8 millimeters

Mathematics I Formula Sheet

Below are the formulas you may find useful as you work the problems. However, some of the formulas may not be used. You may refer to this page as you take the test.

Area

Rectangle and Parallelogram $A = bh$
Triangle $A = \frac{1}{2}bh$
Circle $A = \pi r^2$
Trapezoid $A = \frac{1}{2}(h)(b_1 + b_2)$

Circumference

$C = \pi d$ $\pi \approx 3.14$

Volume

Rectangular Prism/Cylinder $V = Bh$
Pyramid/Cone $V = \frac{1}{3}Bh$

Surface Area

Rectangular Prism $SA = 2lw + 2wh + 2lh$
Cylinder $SA = 2\pi r^2 + 2\pi rh$

Pythagorean Theorem

$$a^2 + b^2 = c^2$$

Mean Absolute Deviation

$$\frac{\sum_{i=1}^{N} \left| x_i - \overline{x} \right|}{N}$$

the average of the absolute deviations from the mean for a set of data

Expected Value

$$E(x) = \sum_{i=1}^{n} x_i \, p(x_i)$$

the sum of each outcome multiplied by its probability of occurrence

Permutations

$$_nP_r = \frac{n!}{(n-r)!}$$

Combinations

$$_nC_r = \frac{n!}{r!(n-r)!}$$

Interquartile Range

the difference between the first quartile and third quartile of a set of data

Practice Test 1

Part 1

You may use the formula sheet on page 230 as needed.

1. Which is the graph of $f(x) = x$?

(A)

(B)

(C)

(D)

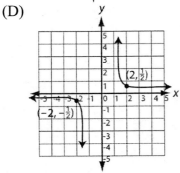

MM1A1b

2. The equation of the function $f(x) = x + 2$ changes to $y = -x - 2$. What happened to the graph of the function?

(A) It reflects across the x-axis.
(B) It reflects across the y-axis.
(C) It shifts down four units.
(D) It does not alter.

MM1A1c

3. Simplify: $f(x) = 4x - \frac{1}{2}x + \sqrt{9}$

(A) $f(x) = 3\frac{1}{2}x + \sqrt{9}$

(B) $f(x) = 4\frac{1}{2}x + 3$

(C) $f(x) = 6\frac{1}{2}$

(D) $f(x) = 3\frac{1}{2}x + 3$

MM1A2a

4. Multiply $(2x)$ and $(x^2 - 3x + 4)$. What is the answer in simplest terms?

(A) $x^3 - 3x^2 + 4x$
(B) $2x^3 - 6x^2 + 8x$
(C) $3x^2 - 2x^6 + 8x$
(D) $2x^3 + 6x^2 - 8x$

MM1A2c

5. What is $(x + 7) \times (x + 4)$ in simplest terms?

(A) $x^2 + 11x + 28$
(B) $x^2 + 3x + 28$
(C) $x^2 - 3x - 28$
(D) $x^2 + 4x + 28$

MM1A2d

6. Divide $3y^3 + 9$ by $y^9 - 9$. What is the answer in simplest terms?

(A) $\dfrac{3}{y^3 - 3}$

(B) $\dfrac{3}{y^3 + 3}$

(C) $\dfrac{3(y^3 + 3)}{y^3 + 3}$

(D) $\dfrac{3(y^3 + 3)}{y^9 - 9}$

MM1A2e

7. Factor: $t^2 - 11t + 30$

(A) $(t + 5)(t - 6)$
(B) $(t - 6)(t + 5)$
(C) $(t - 5)(t - 6)$
(D) $(t - 5)(t + 6)$

MM1A2f

8. What is the intersection point of $f(x) = 2x - 3$ and $g(x) = -x + 9$?

(A) $(4, 5)$
(B) $(1, 8)$
(C) $(6, 9)$
(D) $(12, -3)$

MM1A1i

9. Given the following image, what is x?

(A) 65°
(B) 90°
(C) 180°
(D) 115°

MM1G3a

10. Given the information in the triangle, what is x?

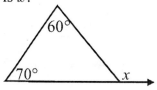

(A) 50°
(B) 130°
(C) 120°
(D) 110°

MM1G3b

11. A _____ is a generalized guess of the outcome of a situation.

(A) counterexample
(B) conjecture
(C) objecture
(D) estimate

MM1G2a

12. Given the two triangles, which theorem for triangles proves these are equivalent?

(A) SSS
(B) SAS
(C) AAS
(D) ASA

MM1G3c

13. What is the sum of the interior angles of the figure?

(A) 180°
(B) 320°
(C) 360°
(D) 540°

MM1G3a

14. What is the point of concurrency where the altitudes of a triangle intersect?

(A) Incenter
(B) Orthocenter
(C) Centroid
(D) Circumcenter

MM1G3e

15. Given the conditional statement, "If an equation is linear, then it has only one solution." What is the contrapositive of that statement?

(A) If an equation is not linear, then it does not have only one solution.
(B) If an equation has only one solution, then it is linear.
(C) If an equation does not have only one solution, then it is not linear.
(D) None of the above

MM1G2b

16. What is the inverse of the statement: "If an animal does not lay eggs, then it is a mammal?"

(A) If an animal does lay eggs, then it not is a mammal.
(B) If an animal is a mammal, then it does not lay eggs.
(C) If an animal is not a mammal, then it does lay eggs.
(D) None of the above

MM1G2b

17. If two girls are trying on shirts and there are ten different colored shirts, how many different choices can the two girls make?

(A) 5
(B) 10
(C) 20
(D) 90

MM1D1b

18. How many permutations can be made out of the letters in MATH?

(A) 4
(B) 16
(C) 20
(D) 24

MM1D1b

19. Jason has a bag of different kinds of chocolates and he picks one and then a second one without replacing the first one. What kind of event is this?

(A) Independent event
(B) Interdependent event
(C) Dependent event
(D) Not enough information

MM1D2b

20. The probability that the Red Sox and the Braves will both win a baseball game in the same day is 20%. The probability that the Braves will win is 55%. What is the probability that the Red Sox will win given that the Braves have already won?

(A) 275%
(B) 36%
(C) 64%
(D) 20%

MM1D2c

21. If you flip a coin 3 times, what is the probability you would get heads every time?

(A) 0.125
(B) 0.25
(C) 0.5
(D) 0.75

MM1D2d

22. The weights of five packages are 5 pounds, 7 pounds, 8 pounds, 5 pounds, and 10 pounds. What are the mean, median, and mode of these weights?

(A) 6.0, 5.5, 7.0
(B) 6.0, 6.5, 7.0
(C) 7.0, 7.0, 5.0
(D) 8.0, 6.0, 5.0

MM1D3a

23. What is the mean absolute deviation of $\{5, 2, 6, 1, 7, 3, 4\}$?

(A) 2

(B) $\dfrac{12}{7}$

(C) $\dfrac{7}{12}$

(D) 9

MM1D4

24. What is the mean absolute deviation of $\{3, 1, 5, 9, 4, 2, 8, 7\}$?

(A) 4.875
(B) 8
(C) 9
(D) 2.375

MM1D4

25. What are the zeros of $y = x^2 - x - 12$?

(A) $x = -4, 3$
(B) $x = 4, -3$
(C) $x = 4$
(D) $x = 3$

MM1A3a

26. In the equation $\frac{1}{2}\sqrt{x} + 1 = 3$, what is x?

(A) 1
(B) 4
(C) 9
(D) 16

MM1A3b

27. Is this function even, odd, or neither?

(A) Even
(B) Odd
(C) Neither
(D) Not enough information

MM1A1h

28. An equation's points are plotted in the table below. According to this table, how many solutions (x-intercepts) does this equation have?

x	$f(x)$
$-\frac{3}{2}$	0
$-\frac{1}{2}$	0
0	5
$\frac{1}{2}$	0
$\frac{3}{2}$	0

(A) 1
(B) 2
(C) 3
(D) 4

MM1A3c

29. In the equation $x + 5 - 2x = 6$, what is x?

(A) -1

(B) 0

(C) 1

(D) 2

MM1A3d

30. What is the distance between $(0, 3)$ and $(-5, 6)$?

(A) 34

(B) $\sqrt{34}$

(C) 36

(D) $\sqrt{36}$

MM1G1a

31. What is the shortest distance between line l and point A? Each unit equals 10 yards.

(A) 5 yards

(B) 10 yards

(C) 25 yards

(D) 50 yards

MM1G1b

32. You are about to get a car for the first time. This means you will also get a licence plate. Each licence plate gets 3 numbers, then 3 letters. The numbers can be the digits from 0 to 9, and the letters can be any letter from A to Z. If the digits cannot be repeated and the letters cannot be repeated in the licence plate, how many possible combinations of the licence plates are there?

(A) 17,576,000

(B) 108

(C) 11,232,000

(D) 27,000

MM1D1a

33. Use the points $A = (-4, 2)$, $B = (3, 2)$, $C = (3, 5)$ and the Pythagorean Theorem to find the distance between A and C. What is the distance?

(A) 25

(B) 5

(C) 75

(D) $\sqrt{58}$

MM1G1d

34. Graph the points $(0, 0)$, $(6, 3)$, $(3, 3)$, $(9, 0)$ to find out what shape they create. What shape do they create?

(A) a trapezoid

(B) a rectangle

(C) a rhombus

(D) a parallelogram

MM1G1e

35. Tina has a bag of jelly beans of four flavors. There are 6 licorice, 4 grape, 3 sour apple, and 5 mint jelly beans in the bag. Tina prefers the fruit-flavored jelly beans. If she selects two jelly beans at random from the bag, what is the probability that the two jelly beans will be grape?

(A) $\dfrac{2}{51}$

(B) $\dfrac{61}{153}$

(C) $\dfrac{4}{9}$

(D) $\dfrac{4}{81}$

MM1D2a

36. Samantha is a student at Etowah High School. Look at the following histograms. The first histogram displays the number of sodas consumed each day by each student in Samantha's math class. The second histogram displays the number of sodas consumed each day by each of the students in the ninth-grade class at her school.

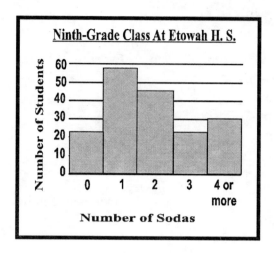

What is the approximate difference in the mode of Samantha's math class and the mode of ninth-grade class at Etowah High School?

(A) 0

(B) 1

(C) 2

(D) 3

MM1D3b

Part 2

37. Which is the graph of $f(x) = \sqrt{x}$?

(A)

(B)

(C)

(D)

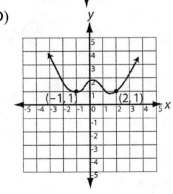

MM1A1b

38. $f(x) = \frac{1}{4}x + \frac{1}{2}$; find $f(12)$.

(A) 4.5
(B) 4
(C) 3.5
(D) 3

MM1A1a

39. The equations $y = 2$ and $y = -2$ illustrate a reflection where?

(A) over the x-axis
(B) over the y-axis
(C) around the origin
(D) There is no reflection.

MM1A1c

40. What is the minimum of $f(x) = x^2 - 1$?

(A) $(0, -1)$
(B) $(0, 0)$
(C) $(0, 1)$
(D) $(0, 2)$

MM1A1d

41. What is the pattern in the following sequence $\{1, 4, 9, 16, 25\}$?

(A) $n + 3$
(B) n^2
(C) $n^2 + 1$
(D) $n + 5(n - 2)$

MM1A1f

42. What is the rate of change of $f(x) = 3x$?

(A) 1
(B) 3
(C) $\frac{1}{3}$
(D) -3

MM1A1g

43. What is $\dfrac{\sqrt{8}}{\sqrt{2}}$ in simplest terms?

(A) $\sqrt{4}$

(B) 4

(C) $\sqrt{2}$

(D) 2

44. Subtract $\dfrac{9}{x+3}$ from $\dfrac{x^2}{5}$. What is the answer in simplest terms?

(A) $x^2 - 9$

(B) $\dfrac{x^3 + 3x^2 - 45}{5x - 15}$

(C) $\dfrac{x^2 - 9}{5x + 15}$

(D) $\dfrac{x^3 + 3x^2 - 45}{5x + 15}$

45. What is the volume of a cube with side length $(x + 2)$?

(A) $x^2 + 4x + 4$

(B) $x^3 + x^2 + 4x + 4$

(C) $x^3 + 6x^2 + 12x + 8$

(D) $x^3 + 12x^2 + 6x + 8$

46. Solve for x: $\dfrac{3x + 2}{7} = -1$

(A) $x = 3$

(B) $x = -3$

(C) $x = -\dfrac{5}{3}$

(D) $x = \dfrac{5}{3}$

47. What is the sum of the angles of a trapezoid?

(A) $180°$

(B) $360°$

(C) $540°$

(D) $720°$

48. In order to form a triangle, what must the third side equal if the first two sides are 1 and 24?

(A) 3

(B) 11

(C) 24

(D) 42

49. Given the two triangles, which theorem for triangles proves these are equivalent?

(A) SSS

(B) SAS

(C) AAS

(D) ASA

50. Which quadrilateral has only one pair of parallel sides?

(A) Square

(B) Rhombus

(C) Parallelogram

(D) Trapezoid

51. What is the name of the center of a triangle that is also the center of a circumscribed circle?

(A) Incenter

(B) Orthocenter

(C) Centroid

(D) Circumcenter

MM1G3e

52. A _____ proves conjectures to be wrong.

(A) counterexample

(B) diverse theory

(C) objecture

(D) inverse

MM1G2a

53. Given the conditional statement: p → q. The statement p → ~q is the _____.

(A) inverse

(B) converse

(C) contrapositive

(D) None of the above

MM1G2b

54. If Steven rolls two dice, what is the probability that the dice total 7 or 11?

(A) $\frac{2}{9}$

(B) $\frac{2}{12}$

(C) $\frac{1}{6}$

(D) $\frac{1}{3}$

MM1D2a

55. How many choices of ice cream sundaes are there if you can pick only one flavor out of 8 and 1 topping out of 6?

(A) 14

(B) 24

(C) 36

(D) 48

MM1D1b

56. Karen has a bag of 50 marbles, 40 are blue and ten are red, every time she draws a blue marble, she _____ the chances of drawing a red marble.

(A) increases

(B) decreases

(C) eliminates

(D) doesn't change

MM1D2b

57. A 2003 survey revealed that 84% of homes have a garage. 65% have a garage and a backyard. What is the probability that a house has a backyard given that it has a garage?

(A) 84%

(B) 65%

(C) 80%

(D) 77%

MM1D2c

58. Daysi has a box with ten balls numbered from 0–9. What is the probability that if she picked out a numbered ball 3 times, she would get the number 7 each time?

(A) $\frac{1}{10}$

(B) $\frac{1}{100}$

(C) $\frac{3}{10}$

(D) $\frac{1}{1000}$

MM1D2d

59. Which set of numbers has the greatest range?

(A) $\{95, 86, 78, 62\}$
(B) $\{90, 65, 83, 59\}$
(C) $\{32, 29, 44, 56\}$
(D) $\{29, 35, 49, 51\}$

MM1D3a

60. In a survey of 125 randomly selected voters, 80 said they would vote for Steven Gillmor. If 20,000 people in the district vote in the election, approximately how many would be expected to vote for Steven Gillmor?

(A) 3,125
(B) 12,800
(C) 14,875
(D) 16,000

MM1D3c

61. In the equation $10 + \sqrt{x} = 20$, what is x?

(A) 100
(B) 50
(C) 10
(D) 5

MM1A3b

62. According to the graph, what are the solution(s) (zeros) to the equation?

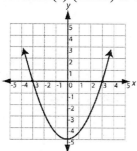

(A) $x = -2, 2$
(B) $x = -3, 3$
(C) $x = -1, 1$
(D) $x = 0$

MM1A3c

63. What is the distance between $(-1, -1)$ and $(0, 5)$?

(A) 6
(B) $\sqrt{36}$
(C) 7
(D) $\sqrt{37}$

MM1G1a

64. To find the shortest distance between a point and a line you must the length of what?

(A) the perpendicular line segment
(B) the parallel line segment
(C) the congruent line segment
(D) the collinear line segment

MM1G1b

65. What is the midpoint of $(6.4, 3)$ and $(-10.7, 4)$?

(A) $(3, 2)$
(B) $(-2.15, 3.5)$
(C) $(-2, 3)$
(D) $(2, -3)$

MM1G1c

66. What is the midpoint of $(-3, 4)$ and $(2, 1)$?

(A) $\left(-\frac{1}{2}, 2\frac{1}{2}\right)$
(B) $(-1, 2)$
(C) $(1, -2)$
(D) $(1, 2)$

MM1G1c

67. Use the points $A = (-4, -4)$, $B = (-4, 2)$, $C = (3, 2)$ and the Pythagorean Theorem to find the distance between A and C. What is the distance?

(A) 100
(B) $\sqrt{85}$
(C) 7
(D) $\sqrt{110}$

MM1G1d

68. What is the interquartile range of the data below?

{1, 1, 2, 3, 7, 8, 10}

(A) 9
(B) 6
(C) 7
(D) 1

MM1D3a

69. Graph the points $(0,0)$, $(1,2)$, $(4,0)$, $(5,2)$ to find out what shape they create. What shape do they create?

(A) a square
(B) a rectangle
(C) a rhombus
(D) a parallelogram

MM1G1e

70. Look at the two figures below. The figure on the left represents the after school activities of one Georgia High School. The figure on the right represents the after school activities for ALL Georgia High Schools. What is the ratio of the children who play sports from one high school to all high schools?

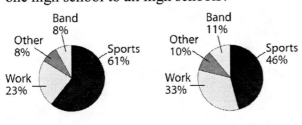

(A) 60 : 40
(B) 61 : 46
(C) 3 : 2
(D) 39 : 54

MM1D3b

71. You are about to get a car for the first time. This means you will also get a licence plate. Each licence plate gets 3 numbers, then 3 letters. The numbers can be the digits from 0 to 9, and the letters can be any letter from A to Z. There are no restrictions on what numbers you can get, but you cannot have any repeating letters. Based on this information, how many possible combinations of the licence plates are there?

(A) 11,232,000
(B) 15,600,000
(C) 17,576,000
(D) 12,654,720

MM1D1a

72. Jon and Nick took 5 tests in their math class this semester. They recorded their scores for each test in the chart below.

Test	Jon	Nick
1	92	84
2	85	89
3	90	96
4	78	84
5	95	92

Who had the greater mean deviation for his score?

(A) Jon
(B) Nick
(C) Neither, they had the same mean deviation.
(D) This is not possible given the information provided.

MM1D4

Practice Test 2

Part 1

You may use the formula sheet on page 230 as needed.

1. $f(x) = \frac{1}{2}x + 7$; find $f(8)$.

    (A) 15

    (B) 4

    (C) 11

    (D) 18

MM1A1a

2. According to the graph, as the x value increases, the y value _____.

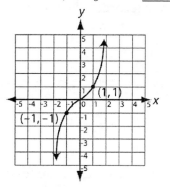

    (A) increases

    (B) decreases

    (C) gets shorter

    (D) gets wider

MM1A1e

3. What is $x + 2 + \sqrt{4}$ in simplest terms?

    (A) $2x + \sqrt{4}$

    (B) $x + 2\sqrt{4}$

    (C) $x + 4$

    (D) $x + 2 + 2$

MM1A2a

4. What is $bc^2 - 8bc - 2b^2c^2$ divided by $2bc$?

    (A) $\frac{c}{2} - 4 - bc$

    (B) $-\frac{c}{2} + 4 + bc$

    (C) $-\frac{2}{c} + 4 + \frac{1}{bc}$

    (D) $\frac{2}{c} - 4 - \frac{1}{bc}$

MM1A2c

5. After multiplication, what is $(x - 1)(x - 3)$ in simplest terms?

    (A) $x^2 - 4x + 3$

    (B) $x^2 - 4x - 3$

    (C) $x^2 + 3x - 4$

    (D) $x^2 - 3x + 4$

MM1A2d

6. What is $64c^2 - 25$ divided by $8c^2 - 5$?

    (A) $(8c - 5)(8c + 5)$

    (B) $\dfrac{(8c - 5)(8c + 5)}{8c - 5}$

    (C) $8c + 5$

    (D) $\dfrac{(8c - 5)(8c + 5)}{8c^2 - 5}$

MM1A2e

7. Using greatest common factors, what are the factors of $20y^3 + 30y^5$?

(A) $5y^3(4 + 6y^2)$
(B) $10y^3(2 + 3y^2)$
(C) $10y^3(4 + 6y^2)$
(D) $10y^5(2 + 3y^2)$

MM1A2f

8. What is the value of x in the equation $-\sqrt{x} + 10 = 7$?

(A) $x = 289$
(B) $x = -289$
(C) $x = 9$
(D) $x = -9$

MM1A3b

9. What is the sum of the interior angles of a square?

(A) $90°$
(B) $180°$
(C) $360°$
(D) $540°$

MM1G3a

10. Using the diagram, which angle has the largest measure?

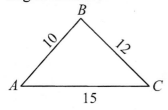

(A) $\angle A$
(B) $\angle B$
(C) $\angle C$
(D) Not enough information

MM1G3b

11. How many different ways can you arrange 5 different pictures on a bookshelf?

(A) 5
(B) 20
(C) 60
(D) 120

MM1D1b

12. Given the two triangles, which theorem for triangles proves these are equivalent?

(A) SSS
(B) SAS
(C) HL
(D) SSA

MM1G3c

13. Which quadrilateral does the figure represent?

(A) Rectangle
(B) Square
(C) Rhombus
(D) Triangle

MM1G3d

14. In a right triangle, which point of concurrency lies on the $90°$ angle?

(A) incenter
(B) centroid
(C) orthocenter
(D) circumcenter

MM1G3e

15. If I said 60 is divisible by all numbers less than ten, to prove me wrong, you would give a(n) _____.

(A) inverse

(B) counterexample

(C) conjecture

(D) converse

MM1G2a

16. Given the statement, "If two triangles are congruent, then they are similar." What is the inverse of that statement?

(A) If two triangles are not congruent, then they are not similar.

(B) If two triangles are similar, then they are congruent.

(C) If two triangles are not similar, then they are not congruent.

(D) None of the above

MM1G2b

17. Given the conditional statement p → q, what does q → p represent?

(A) Inverse

(B) Converse

(C) Contrapositive

(D) None of the above

MM1G2b

18. Which equation represents the number of one thing OR another?

(A) $p(A \text{ or } B) = p(A) + p(B) - p(A \cap B)$

(B) $p(A \text{ or } B) = p(A) + p(B) + p(A \cap B)$

(C) $p(A \text{ or } B) = p(A) + p(B) - p(A \cup B)$

(D) $p(A \text{ or } B) = p(A) + p(B) + p(A \cup B)$

MM1D1a

19. Which equation represents the combination formula?

(A) $_nC_r = \dfrac{n}{n-r}$

(B) $_nC_r = \dfrac{n}{(n-r)r}$

(C) $_nC_r = \dfrac{n!}{(n-r)!r!}$

(D) $_nC_r = \dfrac{n!}{(n-r)!}$

MM1D1b

20. Cindy has a bag of 20 marbles. There are 10 red marbles, 7 black marbles, and 3 blue marbles in the bag. What is the probability that Cindy will draw a red, then a black marble (without replacement)?

(A) $\frac{17}{20}$

(B) $\frac{7}{38}$

(C) $\frac{7}{40}$

(D) $\frac{17}{40}$

MM1D2b

21. A pair of dice is rolled. What is the probability that the sum of the two dice will be greater than four given that the first die rolled is one?

(A) 0.083

(B) 0.167

(C) 0.5

(D) 0.33

MM1D2c

22. What is the mean absolute deviation of $\{9, 4, 8, 3, 6\}$?

(A) 1

(B) 1.5

(C) 2

(D) 2.5

MM1D4

23. A box of a dozen donuts has 3 lemon cream-filled, 5 chocolate cream-filled, and 4 vanilla cream-filled. If the donuts look identical, what is the probability of picking a lemon cream-filled?

(A) $\frac{3}{10}$

(B) $\frac{3}{5}$

(C) $\frac{3}{11}$

(D) $\frac{1}{4}$

MM1D2d

24. Rachel kept track of how many scoops she sold of the five most popular flavors in her ice cream shop.

Flavor	Scoops
Vanilla Bean	30
Chunky Chocolate	36
Strawberry Coconut	44
Chocolate Peanut Butter	46
Mint Chocolate Chip	??

How many scoops of mint chocolate chip would need to be sold to make the mean number of the scoops sold 38?

(A) 34

(B) 35

(C) 36

(D) 38

MM1D3a

25. What is the mean absolute deviation of $\{7, 3, 7, 2, 5, 4, 8, 1\}$?

(A) 2

(B) 2.125

(C) 3

(D) 3.125

MM1D4

26. What are the zeros of $y = x^2 + 5x - 6$?

(A) $x = 6, 1$

(B) $x = -6, -1$

(C) $x = -6, 1$

(D) $x = 6, -1$

MM1A3a

27. Is the equation represented by this graph even, odd, or neither?

(A) Even

(B) Odd

(C) Neither

(D) Not enough information

MM1A1h

28. Given the graph, what is the solution (x-intercept) to the equation of the graph?

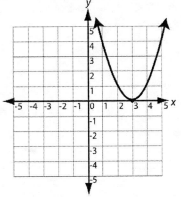

(A) $x = 3$

(B) $x = 0$

(C) $x = 2$

(D) $x = 2.8$

MM1A3c

29. Using the intersection points of the graphs of the equations, what are the solutions of $f(x) = x^2 - 7x - 8$ and $g(x) = 7x + 7$?

(A) $x = 15, -1$
(B) $x = -15, 1$
(C) $x = 7, 8$
(D) $x = -8, 7$

MM1A1i

30. In the equation $2x - 3 = x + 4$, what is x?

(A) $\dfrac{2}{3}$

(B) $-\dfrac{2}{3}$

(C) 7

(D) $\dfrac{3}{4}$

MM1A3d

31. What is the distance between $(-1, 1)$ and $(3, -4)$?

(A) 41
(B) $\sqrt{41}$
(C) 40
(D) $\sqrt{40}$

MM1G1a

32. What is the shortest distance between $y = -1$ and $(1, 2)$?

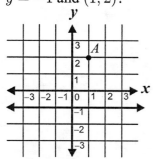

(A) 1 unit
(B) 2 units
(C) 3 units
(D) 4 units

MM1G1b

33. Use the points $A = (-4, -1)$, $B = (-4, -4)$, $C = (-1, -1)$ and the Pythagorean Theorem to find the distance between C and B. What is the distance?

(A) $\sqrt{18}$

(B) 18

(C) $\sqrt{81}$

(D) 81

MM1G1d

34. What kind of triangle do the points $(-3, 0)$, $(3, 0)$, $(0, 3)$ make?

(A) Scalene

(B) Equilateral

(C) Isosceles

(D) They don't form a triangle.

MM1G1e

35. You have a standard deck of 52 cards that has 4 each of 13 kinds of cards. If you draw one card from the deck, what is the probability that the card will be either a Jack or a nine?

(A) $\dfrac{2}{13}$

(B) $\dfrac{1}{13}$

(C) $\dfrac{1}{26}$

(D) $\dfrac{3}{26}$

MM1D2a

36. Samantha is a student at Etowah High School. Look at the following histograms. The first histogram displays the number of sodas consumed each day by each student in Samantha's math class. The second histogram displays the number of sodas consumed each day by each of the students in the ninth-grade class at her school.

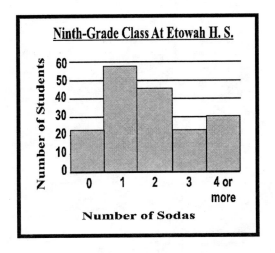

What is the difference in the median of Samantha's math class and the median of ninth-grade class at Etowah High School?

(A) 0

(B) 1

(C) 2

(D) 3

MM1D3b

Part 2

37. Which is the graph of $f(x) = |x|$?

(A)

(B)

(C)

(D)

MM1A1b

38. The graph of $y = 6x + 1$ is reflected over the y-axis. What is the new equation of the graph?

(A) $y = -6x + 1$
(B) $y = -6x - 1$
(C) $y = 6x - 1$
(D) $y = \frac{1}{6}x + 1$

MM1A1c

39. What is the pattern in the following sequence $\{-2, -4, -6, -8\}$?

(A) $-2n$
(B) $2n$
(C) $-\frac{1}{2}n$
(D) $2n + 1$

MM1A1f

40. In the equation $f(x) = x^2 + x + 1$, the rate of change is _____.

(A) constant
(B) variable
(C) nonexistent
(D) not enough information

MM1A1g

41. What is $\dfrac{\sqrt{2}}{\sqrt{6}}$ in simplest terms?

(A) $\dfrac{\sqrt{2} \times \sqrt{6}}{6}$

(B) $\dfrac{\sqrt{12}}{6}$

(C) $\dfrac{2\sqrt{3}}{6}$

(D) $\dfrac{\sqrt{3}}{3}$

MM1A2b

42. After multiplying, what is $2x(x^2 + 3x - 7)$ in simplest terms?

   (A) $2x^3 + 6x^2 - 14x$
   (B) $2x^3 + 6x^2 + 14x$
   (C) $2x^2 + 6x - 14$
   (D) $2x^3 + 3x^2 - 14x$

MM1A2c

43. What is the sum of $\dfrac{5}{y^2}$ and $\dfrac{6}{y^2}$?

   (A) $\dfrac{11}{2y^2}$

   (B) $\dfrac{5+6}{y^2}$

   (C) $\dfrac{11}{y^2}$

   (D) $\dfrac{5}{y^2} + \dfrac{6}{y^2}$

MM1A2e

44. What is the volume of a rectangular prism with length $(x + 1)$, width $(2x - 3)$, and height (x^2)?

   (A) $2x^3 - x - 3$
   (B) $2x^4 - x^3 - 3x^2$
   (C) $2x^4 + x^3 + 3x^2$
   (D) $2x^4 + x^3 - 3x^2$

MM1A2g

45. Using the diagram, what does E equal?

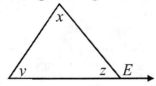

   (A) $x + y$
   (B) $y + z$
   (C) $z + x$
   (D) $x + y + z$

MM1G3b

46. Given the two triangles, which theorem for triangles proves these are equivalent?

   (A) SSS
   (B) AAS
   (C) ASA
   (D) SAS

MM1G3c

47. Which quadrilateral does the figure represent?

   (A) Rectangle
   (B) Square
   (C) Rhombus
   (D) Triangle

MM1G3d

48. The _____ bisects the medians.

   (A) incenter
   (B) centroid
   (C) orthocenter
   (D) circumcenter

MM1G3e

49. A(n) _____ negates the hypothesis and the conclusion of a statement.

   (A) conjecture
   (B) counterexample
   (C) converse
   (D) inverse

MM1G2a

50. Given the conditional statement: p → q, what does ~p → ~q represent?

   (A) Inverse
   (B) Converse
   (C) Contrapositive
   (D) None of the above

MM1G2b

51. If Erin picks up a piece of ice and notices that is cold, then says, "All ice is cold," this is an example of

   (A) deductive reasoning.
   (B) inductive reasoning.
   (C) deduction.
   (D) conjection.

MM1G2a

52. When Katie walked into the ice cream shop, she counted the number of flavors (3) and the number of toppings (8). Assuming she can only get one topping and one flavor, how many choices of combinations does Katie have?

   (A) 11
   (B) 16
   (C) 24
   (D) 32

MM1D1b

53. What is the formula for mutually exclusive events?

   (A) $n(A \cup B) = n(A) + n(B) - n(A \cap B)$

   (B) $\dfrac{n!}{(n-r)!r!}$

   (C) $\dfrac{2pk}{4r}$

   (D) $P(A \text{ or } B) = P(A) + P(B)$

MM1D2a

54. The weights of 8 puppies are 10 pounds, 10 pounds, 11 pounds, 8 pounds, 11 pounds, 13 pounds, 6 pounds, and 11 pounds. What are the mean, median, and mode of these weights?

   (A) 10, 10, 10

   (B) 10, 10.5, 11

   (C) 10, 11, 11

   (D) 10, 10, 11

MM1D3a

55. Blake has a bag of 35 marbles. There are 10 white marbles, 15 yellow marbles, and 10 blue marbles in the bag. What is the probability that Blake will draw a white, then a yellow marble (without replacement)?

   (A) $\dfrac{35}{69}$

   (B) $\dfrac{15}{69}$

   (C) $\dfrac{30}{119}$

   (D) $\dfrac{15}{119}$

MM1D2b

56. A pair of dice is rolled. What is the probability that the sum of the dice is less than 10, given that the first die is four?

   (A) $\dfrac{5}{36}$

   (B) $\dfrac{5}{6}$

   (C) $\dfrac{1}{6}$

   (D) 1

MM1D2c

57. If you have a normal set of dice, what is the probability that if you roll it 3 times, you will get the number 2 each time?

(A) $\frac{1}{6}$

(B) $\frac{1}{36}$

(C) $\frac{1}{216}$

(D) $\frac{1}{360}$

MM1D2d

58. Tom's school was considering making uniforms mandatory starting with the next school year. Tom hated the idea and wanted to do his own survey to see if parents were really in favor of it. He considered 4 places to conduct his survey. Which would give the most valid results?

(A) He would stop people at random walking through the mall.
(B) He would survey parents in the car pool lanes picking up students after school.
(C) He would survey the teachers after school.
(D) He would survey the students in his biology class to ask what their parents thought.

MM1D3c

59. What are the zeros of $y = x^2 - 2x - 3$?

(A) $x = 3, 1$
(B) $x = -3, 1$
(C) $x = -3, -1$
(D) $x = 3, -1$

MM1A3a

60. In the equation $3\sqrt{x} - 1 = 17$, what is x?

(A) 6
(B) 36
(C) 18
(D) 46

MM1A3b

61. The points of the graph of an equation are plotted in the table below. Given this table, what are the solution(s) (x-intercepts) to the equation?

x	0	1	2	3	4	5	6
$f(x)$	16	9	4	1	0	1	4

(A) $x = 4$
(B) $x = 16$
(C) $x = 6, 2$
(D) $x = 3, 5$

MM1A3c

62. In the equation $x^2 + 4x = 7x + 18$, what is x?

(A) $x = 6, 3$
(B) $x = 9, -2$
(C) $x = -6, 3$
(D) $x = 6, -3$

MM1A3d

63. What is the distance between $(1, 7)$ and $(-1, -1)$?

(A) $\sqrt{69}$
(B) $\sqrt{68}$
(C) 8
(D) $\sqrt{67}$

MM1G1a

64. What is the shortest distance between $x = -9$ and $(4, 4)$?

(A) 13 units
(B) 9 units
(C) 5 units
(D) cannot be determined

MM1G1b

65. What is the midpoint of $(5, 4)$ and $(-3, 4)$?

(A) $(0, 4)$

(B) $(-1, 4)$

(C) $(4, 1)$

(D) $(1, 4)$

MM1G1c

66. Use the points $A = (1, -1)$, $B = (3, -1)$, $C = (3, 2)$ and the Pythagorean Theorem to find the distance between A and C. What is the distance?

(A) 9

(B) 4

(C) 13

(D) $\sqrt{13}$

MM1G1d

67. Use the points $A = (3, 2)$, $B = (3, -1)$, $C = (-4, -1)$ and the Pythagorean Theorem to find the distance between A and C. What is the distance?

(A) 3

(B) 9

(C) $\sqrt{10}$

(D) $\sqrt{58}$

MM1G1d

68. What shape do the points $(-4, 0)$, $(-4, 2)$, $(4, 0)$, $(4, 2)$ make?

(A) Rectangle

(B) Square

(C) Parallelogram

(D) Rhombus

MM1G1e

69. The figure on the left depicts sports played by seniors in one high school. The figure on the right depicts sports played by seniors in 100 different high schools. Would figure 1 be a decent representation of all high schools?

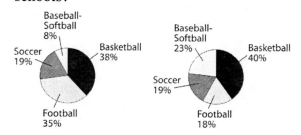

(A) Yes, the proportion of varying sports is about the same.

(B) Yes, because there are a lot of seniors at one high school.

(C) No, the proportions of the sports are not similar.

(D) No, this doesn't provide us with a properly diverse sample.

MM1D3b

70. Cynthia and Robert work at the local video store in their town. The table below shows the amount of hours they worked each week during the month of August.

Week	Cynthia's Hours	Robert's Hours
Week 1	23	29
Week 2	35	31
Week 3	25	30
Week 4	32	28

What is the difference in the median of Cynthia's hours and the median of Robert's hours?

(A) 0.75

(B) 1

(C) 3.25

(D) 9

MM1D3a

71. Nicole is in charge of buying food for a political campaign picnic lunch in City Center Park. In order to determine which foods people would prefer, Nicole conducted a survey of food preferences in front of a small health-food store near the park. The results indicated that most people prefer fruits and vegetables.

Which of the following would be one way that you could improve upon the accuracy of this survey?

(A) She could improve her survey by taking a sample of what types of food her family would prefer.
(B) She could improve her survey by taking a random sample for the population of people that do not like politics.
(C) She could not improve her survey because she picked the best way to survey people.
(D) She could improve her survey by taking a random sample for the population of people who might come to the lunch.

MM1D3c

72. Use number set A to answer the question.

$$A = \{-11, -7, -3, -1, 0, 2, 4, 8, 10, 16\}$$

How many ways can you choose a negative or even number from A?

(A) 11
(B) 10
(C) 9
(D) 7

MM1D1a

Index